ASCE STANDARD ASCE/SEI 24-05

American Society of Civil Engineers

Flood Resistant Design and Construction

This document uses both the International System of Units (SI) and customary units.

ASCE

SEI
Structural Engineering Institute
of the American Society of Civil Engineers

Published by the American Society of Civil Engineers

Published by American Society of Civil Engineers
1801 Alexander Bell Drive
Reston, Virginia 20191
www.pubs.asce.org

Cataloging-in-Publication Data on file with the Library of Congress

Any statements expressed in these materials are those of the individual authors and do not necessarily represent the views of ASCE, which takes no responsibility for any statement made herein. No reference made in this publication to any specific method, product, process, or service constitutes or implies an endorsement, recommendation, or warranty thereof by ASCE.

ASCE makes no representation or warranty of any kind, whether express or implied, concerning the accuracy, completeness, suitability, or utility of any information, apparatus, product, or process discussed in this publication, and assumes no liability therefore. This information should not be used without first securing competent advice with respect to its suitability for any general or specific application. Anyone utilizing this information assumes all liability arising from such use, including but not limited to infringement of any patent or patents.

ASCE and American Society of Civil Engineers—
Registered in U.S. Patent and Trademark Office.

ISBN 0-7844-0818-1
Manufactured in the United States of America.

STANDARDS

In April 1980, the Board of Direction approved ASCE Rules for Standards Committees to govern the writing and maintenance of standards developed by the Society. All such standards are developed by a consensus standards process managed by the Codes and Standards Activities Committee. The consensus process includes balloting by the Balanced Standards Committee, which is composed of Society members and nonmembers, balloting by the membership of ASCE as a whole, and balloting by the public. All standards are updated or reaffirmed by the same process at intervals not exceeding 5 years.

The following Standards have been issued:

ANSI/ASCE 1-82 N-725 Guideline for Design and Analysis of Nuclear Safety Related Earth Structures
ANSI/ASCE 2-91 Measurement of Oxygen Transfer in Clean Water
ANSI/ASCE 3-91 Standard for the Structural Design of Composite Slabs and ANSI/ASCE 9-91 Standard Practice for the Construction and Inspection of Composite Slabs
ASCE 4-98 Seismic Analysis of Safety-Related Nuclear Structures
Building Code Requirements for Masonry Structures (ACI 530-02/ASCE 5-02/TMS 402-02) and Specifications for Masonry Structures (ACI 530.1-02/ASCE 6-02/TMS 602-02)
ASCE/SEI 7-05 Minimum Design Loads for Buildings and Other Structures
ANSI/ASCE 8-90 Standard Specification for the Design of Cold-Formed Stainless Steel Structural Members
ANSI/ASCE 9-91 listed with ASCE 3-91
ASCE 10-97 Design of Latticed Steel Transmission Structures
SEI/ASCE 11-99 Guideline for Structural Condition Assessment of Existing Buildings
ASCE 12-05 Guideline for the Design of Urban Subsurface Drainage
ASCE 13-05 Standard Guidelines for Installation of Urban Subsurface Drainage
ASCE 14-05 Standard Guidelines for Operation and Maintenance of Urban Subsurface Drainage
ASCE 15-98 Standard Practice for Direct Design of Buried Precast Concrete Pipe Using Standard Installations (SIDD)
ASCE 16-95 Standard for Load Resistance Factor Design (LRFD) of Engineered Wood Construction
ASCE 17-96 Air-Supported Structures
ASCE 18-96 Standard Guidelines for In-Process Oxygen Transfer Testing
ASCE 19-96 Structural Applications of Steel Cables for Buildings
ASCE 20-96 Standard Guidelines for the Design and Installation of Pile Foundations
ASCE 21-96 Automated People Mover Standards—Part 1
ASCE 21-98 Automated People Mover Standards—Part 2
ASCE 21-00 Automated People Mover Standards—Part 3
SEI/ASCE 23-97 Specification for Structural Steel Beams with Web Openings
ASCE/SEI 24-05 Flood Resistant Design and Construction
ASCE 25-97 Earthquake-Actuated Automatic Gas Shut-Off Devices
ASCE 26-97 Standard Practice for Design of Buried Precast Concrete Box Sections
ASCE 27-00 Standard Practice for Direct Design of Precast Concrete Pipe for Jacking in Trenchless Construction
ASCE 28-00 Standard Practice for Direct Design of Precast Concrete Box Sections for Jacking in Trenchless Construction
SEI/ASCE/SFPE 29-99 Standard Calculation Methods for Structural Fire Protection
SEI/ASCE 30-00 Guideline for Condition Assessment of the Building Envelope
SEI/ASCE 31-03 Seismic Evaluation of Existing Buildings
SEI/ASCE 32-01 Design and Construction of Frost-Protected Shallow Foundations
EWRI/ASCE 33-01 Comprehensive Transboundary International Water Quality Management Agreement
EWRI/ASCE 34-01 Standard Guidelines for Artificial Recharge of Ground Water
EWRI/ASCE 35-01 Guidelines for Quality Assurance of Installed Fine-Pore Aeration Equipment
CI/ASCE 36-01 Standard Construction Guidelines for Microtunneling
SEI/ASCE 37-02 Design Loads on Structures During Construction
CI/ASCE 38-02 Standard Guideline for the Collection and Depiction of Existing Subsurface Utility Data
EWRI/ASCE 39-03 Standard Practice for the Design and Operation of Hail Suppression Projects
ASCE/EWRI 40-03 Regulated Riparian Model Water Code
ASCE/EWRI 42-04 Standard Practice for the Design and Operation of Precipitation Enhancement Projects
ASCE/SEI 43-05 Seismic Design Criteria for Structures, Systems, and Components in Nuclear Facilities
ASCE/EWRI 44-05 Standard Practice for the Design and Operation of Supercooled Fog Dispersal Projects
ASCE/EWRI 45-05 Standard Guidelines for the Design of Urban Stormwater Systems
ASCE/EWRI 46-05 Standard Guidelines for the Installation of Urban Stormwater Systems
ASCE/EWRI 47-05 Standard Guidelines for the Operation and Maintenance of Urban Stormwater Systems

FOREWORD

The material presented in this standard has been prepared in accordance with recognized engineering principles. This standard should not be used without first securing competent advice with respect to its suitability for any given application. The publication of the material contained herein is not intended as a representation or warranty on the part of the American Society of Civil Engineers, or of any other person named herein, that this information is suitable for any general or particular use or promises freedom from infringement of any patent or patents. Anyone making use of this information assumes all liability from such use.

ACKNOWLEDGEMENTS

The American Society of Civil Engineers (ASCE) acknowledges the work of the Flood Resistant Design and Construction Standards Committee of the Codes and Standards Activities Division of the Structural Engineering Institute. This group comprises individuals from many backgrounds including consulting engineering, research, construction industry, education, government, design, and private practice.

This revision of the standard began in 2001 and incorporates information as described in the commentary.

This standard was prepared through the consensus standards process by balloting in compliance with procedures of ASCE's Codes and Standards Activities Committee. Those individuals who serve on the Standards Committee are

Conrad Battreal
William L. Coulbourne, P.E.
Shou Shan Fan, Ph.D.
Daryle L. Fontenot
Kenneth A. Ford
James S. Graham, Jr., P.E.
Michael Graham
Barbara D. Hayes
John L. Ingargiola
Christopher P. Jones, P.E., Vice Chair
Thomas MacAllen
Joseph J. Messersmith, Jr., P.E.
Kim Paarlberg
Spencer M. Rogers, Jr.
Herbert S. Saffir, P.E., L.S.
Edward M. Salsbury, P.E.
Erez Sela, P.E.
Paul R. Tertell
Harry B. Thomas, P.E., Chair
Theodore C. Van Kirk
Robert A. Wessel
Thomas G. Williamson, P.E.

CONTENTS

Commentary

Flood Resistant Design and Construction

1.0 GENERAL

1.1 SCOPE

This standard provides minimum requirements for flood-resistant design and construction of structures that are subject to building code requirements and that are located, in whole or in part, in flood hazard areas. This standard applies to new construction that includes: (a) new structures, including subsequent work to such structures, and (b) work classified as substantial repair or substantial improvement of an existing structure that is not an historic structure (see Fig. 1-1).

The general provisions of this section shall apply to all new construction and substantial improvements in flood hazard areas. In addition to the requirements of this section (see Fig. 1-2):

1. Section 2 shall apply to all new construction and substantial improvements in Flood Hazard Areas and High Risk Flood Hazard Areas except those that are identified as Coastal High Hazard Areas and Coastal A Zones;
2. Section 3 shall apply to all new construction and substantial improvements in High Risk Flood Hazard Areas;
3. Section 4 shall apply to all new construction and substantial improvements in Coastal High Hazard Areas and Coastal A Zones; and
4. Sections 5, 6, 7, 8, and 9 shall apply to all new construction and substantial improvements.

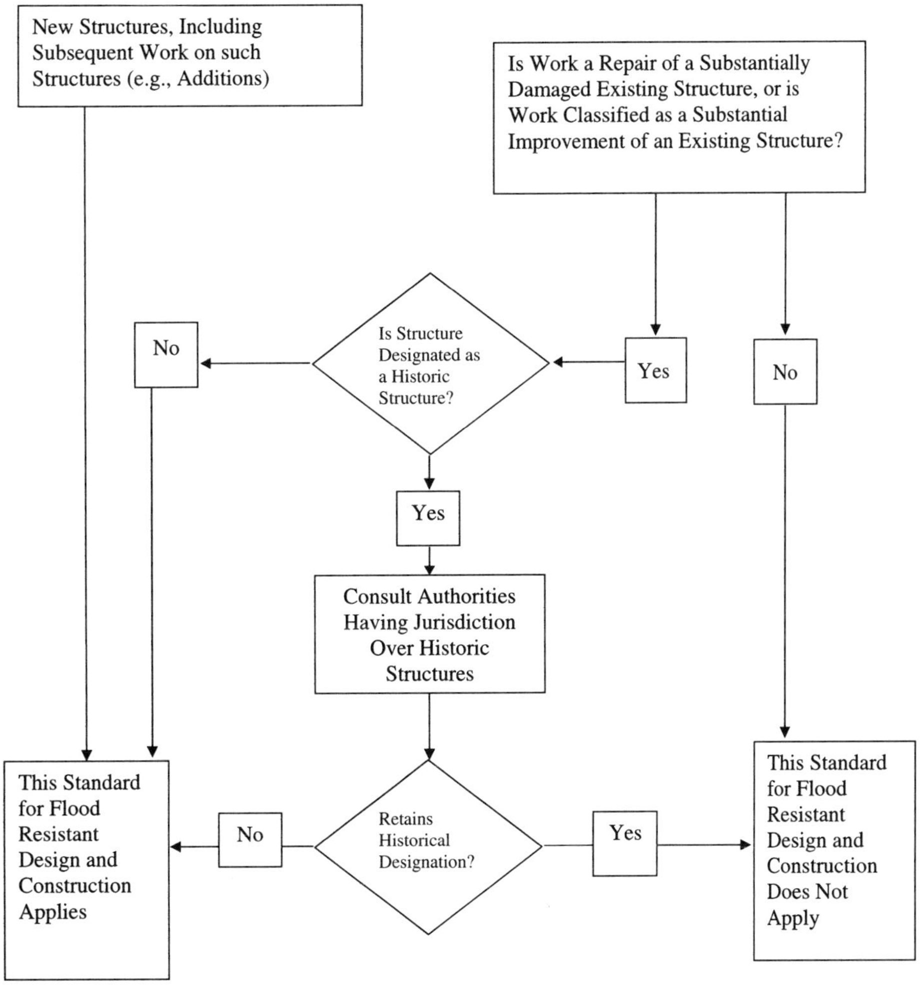

FIGURE 1-1. Illustration of Application of this Standard.

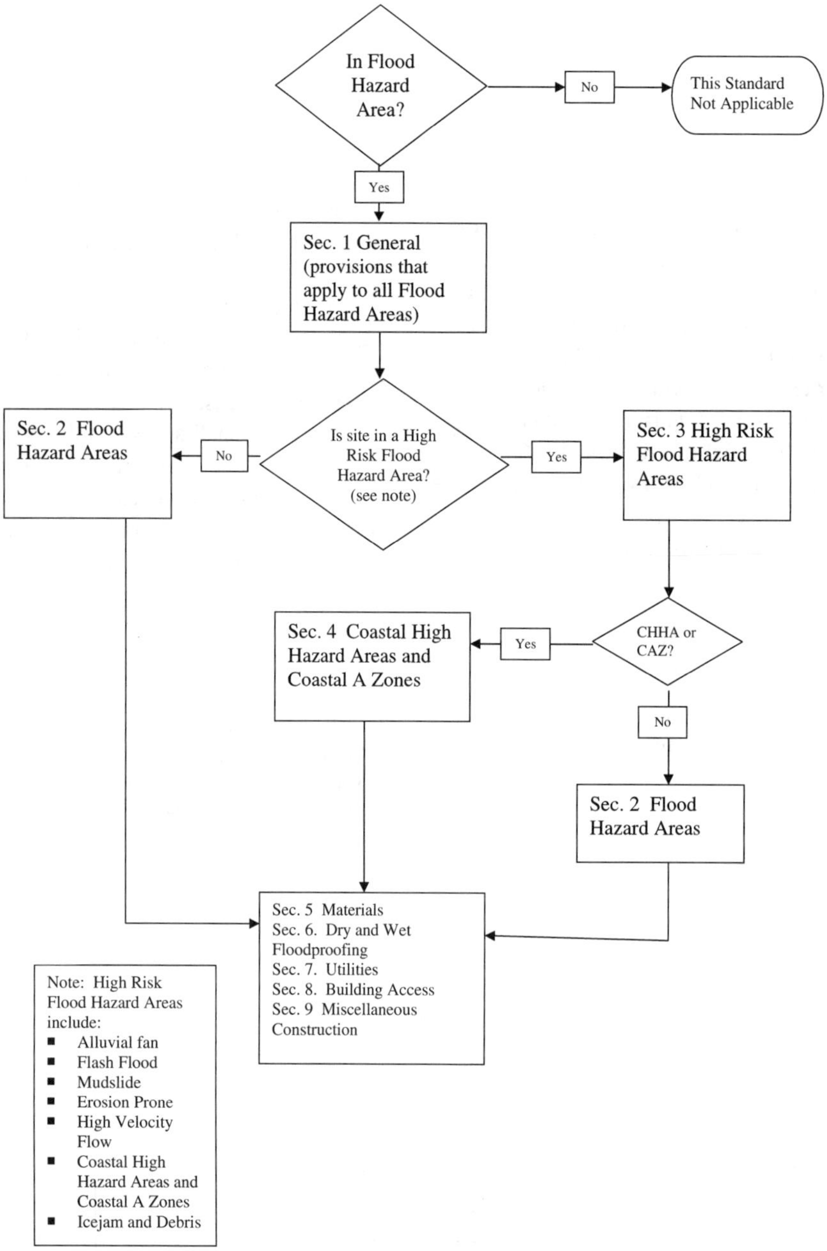

FIGURE 1-2. Application of Sections.

1.2 DEFINITIONS

The following definitions apply to the provisions of the entire standard (italicized words in a definition mean the words are defined in this section):

Alluvial Fan—fan-shaped deposits of sediment eroded from steep slopes and watersheds and deposited on valley floors.

Alluvial Fan Flooding—type of flood hazard that occurs only on *alluvial fans.* Alluvial fan flooding is considered hazardous when designated as a *flood hazard area* on a community's flood hazard map or otherwise legally designated.

Apex—highest point on an *alluvial fan* or similar landform, where the flow is last confined. The apex generally corresponds to the location where the *watershed* erosion ceases and fan sediment deposition commences.

Attendant Utilities and Equipment—utilities, plumbing, HVAC, and related equipment, as well as services associated with new construction.

Authority Having Jurisdiction—organization, community, political subdivision, office, or agency that has adopted this standard under due legislative authority.

Base Flood Elevation (BFE)—elevation of flooding, including wave height, having a 1% chance of being equaled or exceeded in any given year.

Base Flood—flood having a 1% chance of being equaled or exceeded in any given year.

Basement—that portion of a *structure* having its lowest floor below ground level on all sides.

Breakaway Wall—any type of wall subject to flooding that is not required to provide structural support to a building or other structure, and that is designed and constructed such that, under *base flood* or lesser flood conditions, it will collapse in such a way that (1) it allows the free passage of floodwaters; and (2) it does not damage the structure or supporting foundation system.

Bulkhead—wall or structure to retain or prevent sliding or erosion of the land; sometimes used to protect against wave action.

Channel—natural or artificial waterway that periodically or continuously contains moving water.

Coastal A Zone—area within a *special flood hazard area,* landward of a *V Zone* or landward of an open coast without mapped *V Zones.* In a Coastal A Zone, the principal source of flooding must be astronomical tides, storm surges, seiches, or tsunamis, not riverine flooding. During the *base flood* conditions, the potential for breaking *wave heights* shall be greater than or equal to 1.5 ft.

Coastal High Hazard Area (CHHA)—area within a *special flood hazard area* extending from offshore to the inland limit of a primary frontal dune along an open coast and any other area that is subject to high velocity wave action from storms or seismic sources. This area is designated on *FIRM*s as velocity zones V, VO, VE, or V1-30.

Community—any state or area or political subdivision thereof, or any Indian tribe or authorized tribal organization, or Alaska native village or authorized native organization, which has the authority to adopt and enforce this standard for areas within its jurisdiction.

Debris Flow—mass movement of sediment, including boulders, organic materials, and other debris; debris flows typically move in surges and are characterized by a steep frontal wave.

Debris Impact Loads—loads on a *structure* caused by flood-borne debris striking the structure, or a portion thereof, often sudden in nature and large in magnitude.

Design Flood—greater of the following two flood events: (1) the *base flood,* affecting those areas identified as *special flood hazard areas* on the community's *FIRM;* or (2) the flood corresponding to the area designated as a *flood hazard area* on a community's *flood hazard map* or otherwise legally designated.

Design Flood Elevation (DFE)—elevation of the *design flood,* including *wave height,* relative to the datum specified on the community's *flood hazard map.*

Dry Floodproofing—floodproofing method used to render a *structure* envelope *substantially impermeable* to the entrance of floodwater.

Enclosure—confined area below the *DFE,* formed by walls on all sides of the enclosed space.

Eroded Ground Elevation—elevation of ground following *erosion.*

Erodible Soil—soil subject to wearing away and movement due to the effects of wind, water, or other geological processes, during a flood or storm or over a period of years.

Erosion—wearing away of the land surface by detachment and movement of soil and rock fragments, during a flood or storm or over a period of years, through the action of wind, water, or other geologic processes.

Erosion Analysis—analysis of the *erosion* potential of soil or strata, to include the effects of flooding or storm surge, moving water, wave action, and the interaction of water and structural components.

Existing Structure—any *structure* for which the *start of construction* commenced before the effective date of the first floodplain management code, ordinance, or standard adopted by the *authority having jurisdiction.*

Fill—material such as soil, gravel, or crushed stone that is placed in an area to increase ground elevations (see *Structural Fill*).

Flash Flood—*flood* that crests in a short length of time and is often characterized by *high velocity flow;* it often results from heavy rainfall over a localized area, which overflows a confined water course. A flood whose waters rise from within banks to 3 ft or more above banks in less than 2 hours shall be considered a flash flood.

Flood or Flooding—general and temporary condition of partial or complete inundation of normally dry land from (1) the overflow of inland or tidal waters; or (2) the unusual and/or rapid accumulation of runoff or surface waters from any source.

Flood Control Structure—barrier designed and constructed to keep water away from or out of a specified area (see *flood protective works*).

Flood-Damage-Resistant Material—any construction material capable of withstanding direct and *prolonged contact with floodwaters,* without sustaining any damage that requires more than cosmetic repair.

Flood Hazard Area—area subject to flooding during the *design flood.*

Flood Hazard Map—map delineating *flood hazard areas* adopted by the *authority having jurisdiction.*

Flood Hazard Study—study that serves as the technical basis for a *flood hazard map.*

Flood Insurance Rate Map (FIRM)—official map of a community on which the Federal Insurance and Mitigation Administration has delineated both *special flood hazard areas* and the risk premium zones applicable to the community.

Flood Protective Works—barriers designed and constructed to keep water away from or out of a specified area (see *flood control structure; floodwall*).

Flood-Related Erosion—collapse, subsidence, or wearing away of land as a result of the action of flooding, including the effects of storm surge, moving water, and wave action.

Floodplain—any land area, including watercourse, susceptible to partial or complete inundation by water from any source.

Floodproofing—any combination of structural or nonstructural adjustments, changes, or actions that reduce or eliminate flood damage to a *structure,* contents, and *attendant utilities and equipment.*

Floodwall—constructed barrier of *flood-damage-resistant materials* used to keep water away from or out of a specified area.

Floodway—channel and that portion of the *floodplain* reserved to convey the base flood without cumulatively increasing the water surface elevation more than a designated height.

Footing—enlarged base of a foundation, wall, pier, or column, designed to spread the load of the *structure* so that it does not exceed the soil-bearing capacity.

Footprint—horizontal extent of a *structure.*

High Risk Flood Hazard Area—*flood hazard area* where one or more of the following hazards are known to occur: *alluvial fan flooding, flash floods, mudslides, ice jams, high velocity flows, high velocity wave action,* breaking wave heights greater than or equal to 1.5 ft (*Coastal High Hazard Area* and *Coastal A Zone*), or *erosion.*

High Velocity Flow—during *design flood* or lesser conditions, water movement adjacent to *structures* and/or foundations with flow velocities greater than 10 ft/sec.

High Velocity Wave Action—condition where wave heights are greater than or equal to 3.0 ft in height or where wave runup elevations reach 3.0 ft or more above grade.

Highest Adjacent Grade—highest elevation of the natural or regraded ground surface, or *structural fill,* at the location of a *structure.*

Historic Structure—any *structure* that meets one of the following criteria: (1) listed individually in the National Register of Historic Places; (2) certified by the Secretary of the Interior as meeting the requirements for individual listing in the National Register; (3) certified or preliminary determination by the Secretary of the Interior as contributing to the historical significance of a registered historic district or a district preliminarily determined by the Secretary to qualify as a registered historic district; (4) individually listed on a state inventory of historic places, in states with historic preservation programs that have been approved by the Secretary of the Interior; or (5) individually listed on a local inventory of historic places in communities with historic programs certified by an approved state program or by the Secretary of the Interior.

Human Intervention—required presence and active involvement of people to enact a *floodproofing* measure prior to flooding.

Hydrodynamic Loads—loads imposed on an object by water flowing against and around it.

Hydrostatic Loads—loads imposed on an object by a standing mass of water.

Ice jam—accumulation of floating ice fragments that causes the bridging or damming of a channel or stream.

Impact Loads—loads that result from debris, ice, or any object transported by floodwaters striking against *structures* or parts thereof.

Levee—man-made barrier, usually an earthen embankment, designed and constructed in accordance with sound engineering practices, to contain, control, or divert the flow of water so as to provide protection from temporary flooding.

Local Scour—during flood conditions, the removal of material from a localized portion of the channel cross-section or land surface, due to an abrupt change in flow direction or velocity around an object or structural element.

Lowest Floor—lowest floor of the lowest enclosed area, including basement; however, an unfinished or flood-resistant enclosure used solely for parking, building access, or storage shall not be considered the lowest floor provided such enclosure is built as specified in this standard.

Mangrove Stand—assemblage of mangrove trees containing one or more of the following species: black mangrove, red mangrove, white mangrove, or buttonwood.

Mud Flood—hyperconcentrated sediment flow with a sediment concentration less than 45% by volume. Characterized by distinct fluid properties in deformation, particle setting, wave motion, and spreading on a horizontal surface. A mud flood is a turbulent flood phenomenon.

Mudflow—hyperconcentrated sediment flow with a sediment concentration in excess of 45% by volume, characterized by plastic deformation, cohesive characteristics, and lack of wave motion and spreading on a horizontal surface. A mudflow is a viscous flow.

Mudslide—general category of hyperconcentrated sediment flows including *mudflows, mud floods,* and *debris flows.*

NAVD—North American Vertical Datum.

New Construction—new structures including subsequent work to such structures, and work classified as substantial repair or substantial improvements of an existing structure that is not a historical structure.

New Structure—any structure for which *start of construction* commenced after the effective date of the first floodplain management code, ordinance, or standard adopted by the *authority having jurisdiction.*

NGVD—National Geodetic Vertical Datum.

Nonerodible Soil—soil not subject to wearing away or movement due to the effects of wind, water, or other agents during a flood or storm or a period of years.

Nonresidential—any building or structure or portion thereof that is not classified residential.

Obstruction—any object or structural component attached to a *structure* below the *DFE* that can cause an increase in flood elevation, deflect flood waters, or transfer flood loads to any structure.

Pile—structural element that is embedded in soils by drilling, driving, or jetting, so that axial loads in the member are supported through skin friction or end bearing with the soil or rock and lateral loads in the member are supported through side bearing with the soil or rock.

Prolonged Contact with Floodwaters—partial or total inundation by floodwaters for (1) 72 hours for non-coastal *flood hazard areas;* and (2) 12 hours for *Coastal High Hazard Areas* and *Coastal A Zones.*

Rapid Drawdown—rapid lowering of flood elevation, at a rate equal to or exceeding 5 ft/hr.

Rapid Rise—rapid increase in flood elevation, at a rate equal to or exceeding 5 ft/hr.

Residential—(1) buildings and structures and portions thereof where people live or that are used for sleeping purposes on a transient or non-transient basis; (2) residential structures, including but not limited to one- and two-family dwellings, townhouses, condominiums, multifamily dwellings, apartments, congregate residences, boarding houses, lodging houses, rooming houses, hotels, motels, apartment buildings, convents, monasteries, dormitories, fraternity houses, sorority houses, vacation time-share properties; and (3) institutional facilities where people are cared for or live on a 24-hour basis in a supervised environment, including but not limited to board and care facilities, assisted living facilities, halfway houses, group homes, congregate care facilities, social rehabilitation facilities, alcohol and drug centers, convalescent facilities, hospitals, nursing homes, mental hospitals, detoxification facilities, prisons, jails, reformatories, detention centers, correctional centers, and prerelease centers.

Sand Dune—natural or artificial ridges or mounds of sand landward of a beach.

Scour—removal of soil or *fill* material from the channel cross-section or land surface by the flow of flood waters.

Seawall—wall separating land and water areas, primarily designed to prevent *erosion* and other damage due to wave action.

Shear Wall—load bearing or non-load–bearing wall that transfers, by in-plane lateral forces, lateral loads acting on a *structure* to its foundation.

Shield—removable or permanent substantially impermeable protective cover for an opening in a *structure* below the *DFE,* used in *floodproofing* the *structure.*

Special Flood Hazard Area—land in the floodplain subject to a 1% or greater chance of flooding in any given year; area delineated on the *Flood Insurance Rate Map* as Zone A, AE, A1-30, A99, AR, AO, AH, V, VO, VE, or V1-30.

Start of Construction—date the construction permit was issued for *new construction,* provided that actual start of construction commenced within 180 days of the permit date. The actual start means either the first placement of permanent construction of a *structure* on a site, such as the pouring of a slab or *footing,* the installation of *piles,* the construction of columns, or any other work beyond the stage of excavation; or the placement of a manufactured home. Permanent construction does not include land preparation, such as clearing, grading, or filling; nor does it include excavation for a *basement, footings,* piers, or foundation or the erection of temporary forms; nor does it include the installation of accessory structures, such as garages or sheds not occupied as dwelling units or not part of the main structure. For *substantial repair* or *substantial improvement,* the actual start of construction means the first alteration of any wall, ceiling, floor, or other structural part of a *structure,* whether or not that alteration affects the external dimensions of the *structure.*

Stillwater Depth—vertical distance between the ground and the *stillwater elevation.*

Stillwater Elevation—elevation that the surface of the water would assume in the absence of waves referenced to *NAVD, NGVD,* or other datum.

Storage Tank—closed vessel used to store gases or liquids.

Structural Fill—*Fill* compacted to a specified density to provide structural support or protection to a *structure.*

Structure—any building or other *structure,* including gas and liquid storage tanks.

Substantial Damage—damage of any origin sustained by a *structure,* whereby the cost of restoration to its predamage condition equals or exceeds 50% of its predamage market value, or equals or exceeds a smaller percentage established by the *authority having jurisdiction.*

Substantial Improvement—any reconstruction, rehabilitation, addition, or other improvement to a *structure,* the cost of which equals or exceeds 50% of its preimprovement market value, or equals or exceeds a smaller percentage established by the *authority having jurisdiction.*

Substantial Repair—repair of a *substantially damaged* structure to which the requirements of this standard apply.

Substantially Impermeable—use of *flood-damage-resistant materials* and techniques for *dry floodproofing* portions of a *structure,* which result in a space free of through cracks, openings, or other channels that permit unobstructed passage of water and seepage during flooding, and which result in a maximum accumulation of 4 in. of water depth in such space during a period of 24 hours.

V Zone—Velocity Zones V, VO, VE, or V1-30 (See *Coastal High Hazard Area*).

Watershed—topographically defined area drained by a river or stream, or by a system of connecting rivers and streams such that all outflow is discharged through a single outlet.

Wave—ridge, deformation, or undulation of the water surface.

Wave Crest Elevation—elevation of the crest of a *wave,* usually referenced to *NGVD, NAVD,* or another datum.

Wave Height—vertical distance between the crest and the trough of a *wave.*

Wave Loads—loads imparted on a *structure* caused by *waves* striking the *structure,* or a portion thereof.

Wave Runup—rush of *wave* water running up a slope or *structure.*

Wave Runup Elevation—elevation, usually referenced to *NGVD, NAVD,* or another datum, reached by *wave runup.*

Wet Floodproofing—*floodproofing* method that relies on the use of *flood-damage-resistant materials* and construction techniques to minimize flood damages to areas below the *DFE* of a *structure* intentionally allowed to flood (see *floodproofing*).

1.3 IDENTIFICATION OF FLOOD HAZARD AREAS

This standard shall apply to the larger of the two areas listed below:

1. Those lands within a floodplain subject to a 1%, or greater, chance of flooding in any year (i.e., the area subject to flooding during the base flood event); and
2. Those lands designated as a flood hazard area on a community's flood hazard map, or otherwise legally designated.

The flood associated with the governing definition listed above shall be termed the design flood. Design and construction requirements for new construction and substantial improvements shall be dictated by conditions during the design flood.

1.4 IDENTIFICATION OF FLOODPRONE STRUCTURES

1.4.1 General

A determination shall be made as to whether or not a structure lies, in whole or in part, within a flood hazard area following review of flood hazard maps, studies available in public domain, and other information available from the authority having jurisdiction.

1.4.2 Consideration for Flood Protective Works

Dams, levees, floodwalls, diversions, channels, and other flood protective works shall not be considered to provide protection for structures during the design flood, unless those works are shown on the flood hazard map as providing protection during design flood conditions or unless those works are determined by the authority having jurisdiction to provide protection during design flood conditions.

Design of structures behind levees and flood-walls shall consider the adequacy of drainage of rainfall, runoff, and other waters behind the levees and floodwalls.

New construction in proximity to flood protective works shall not

1. Damage, endanger, or otherwise harm the flood protective works;
2. Be in conflict with maintenance and repair operations of the entity responsible for the flood protective works; or
3. Significantly increase the potential for trapping of rainfall, runoff, and other waters behind the flood protective works.

1.4.3 Classification of Structures

Structures shall be classified according to Table 1-1.

TABLE 1-1. Classification of Structures for Flood-Resistant Design and Construction (Classification same as ASCE 7, Ref. [1])

Nature of Occupancy	Category
Buildings and other structures that represent a low hazard to human life in the event of failure including, but not limited to: • Agricultural facilities[a] • Certain temporary facilities • Minor storage facilities[b]	I
All buildings and other structures except those listed in Categories I, III, and IV	II
Buildings and other structures that represent a substantial hazard to human life in the event of failure including, but not limited to: • Buildings and other structures where more than 300 people congregate in one area • Buildings and other structures with day-care facilities with capacity greater than 150 • Buildings and other structures with elementary school or secondary school facilities with capacity greater than 250 • Buildings and other structures with a capacity greater than 500 for colleges or adult education facilities • Health care facilities with a capacity of 50 or more resident patients but not having surgery or emergency treatment facilities • Jails and detention facilities • Power generating stations and other public utility facilities not included in Category IV Buildings and other structures not included in Category IV (including, but not limited to, facilities that manufacture, process, handle, store, use, or dispose of such substances as hazardous fuels, hazardous chemicals, hazardous waste, or explosives) containing sufficient quantities of hazardous materials considered to be dangerous to the public if released. Buildings and other structures containing hazardous materials shall be eligible for classification as Category II structures if it can be demonstrated to the satisfaction of the authority having jurisdiction by a hazard assessment as described in Section 1.5.2[c] that a release of the hazardous material does not pose a threat to the public.	III
Buildings and other structures designated as essential facilities including, but not limited to: • Hospitals and other health care facilities having surgery or emergency treatment facilities • Fire, rescue, ambulance, and police stations and emergency vehicle garages • Designated earthquake, hurricane, or other emergency shelters • Designated emergency preparedness, communication, and operation centers and other facilities required for emergency response • Power generating stations and other public utility facilities required in an emergency • Ancillary structures (including, but not limited to, communication towers, fuel storage tanks, cooling towers, electrical substation structures, fire water storage tanks or other structures housing or supporting water, or other fire-suppression material or equipment) required for operation of Category IV structures during an emergency • Aviation control towers, air traffic control centers, and emergency aircraft hangars • Water storage facilities and pump structures required to maintain water pressure for fire suppression • Buildings and other structures having critical national defense functions Buildings and other structures (including, but not limited to, facilities that manufacture, process, handle, store, use, or dispose of such substances as hazardous fuels, hazardous chemicals, hazardous waste, or explosives) containing extremely hazardous materials where the quantity of the material exceeds a threshold quantity established by the authority having jurisdiction. Buildings and other structures containing extremely hazardous materials shall be eligible for classification as Category II structures if it can be demonstrated to the satisfaction of the authority having jurisdiction by a hazard assessment as described in Section 1.5.2[c] that a release of the extremely hazardous material does not pose a threat to the public. This reduced classification shall not be permitted if the buildings or other structures also function as essential facilities.	IV

[a]Certain agricultural structures may be exempt from some of the provisions of this standard—see Section C1.4.3.
[b]For the purposes of this standard, minor storage facilities do not include commercial storage facilities.
[c]Section 1.5.2 reference is made to ASCE Standard 7-05, not this standard.

1.5 BASIC DESIGN AND CONSTRUCTION REQUIREMENTS

1.5.1 General

New construction and substantial improvements shall be designed, constructed, connected, and anchored to resist flotation, collapse, or permanent lateral movement resulting from the action of hydrostatic, hydrodynamic, wind, and other loads during design flood, or lesser, conditions in accordance with requirements of this standard if specified, or if not specified in this standard then in accordance with requirements approved by the authority having jurisdiction. Design shall include those loads and load combinations described in Section 1.6.

Design and construction in flood hazard areas shall account for each of the following:

1. Elevation of the structure relative to the design flood elevation (DFE);
2. Foundations and geotechnical factors;
3. Damage to the structure up to and during the design flood;
4. Obstructions or enclosures below the DFE;
5. Structural connections;
6. Use of flood-damage-resistant materials;
7. Floodproofing;
8. Utilities;
9. Means of egress; and
10. Adverse impacts to other structures and property.

1.5.2 Elevation Requirements

Structures shall have the lowest floor (including basements) elevated to or above the DFE in conformance with the requirements of the section applicable to the specific flood hazard area.

Enclosed areas that are used solely for parking, building access, or storage are not the lowest floor and shall be allowed below the DFE provided the enclosed areas meet the requirements for enclosed areas applicable to the specific flood hazard area.

Nonresidential structures and nonresidential portions of mixed-use structures shall be allowed to have the lowest floor (including basements) below the DFE provided the structures meet the dry floodproofing requirements in Section 6.

Structures that are used solely for parking, building access, or storage shall be allowed below the DFE provided the structures meet the requirements of Section 9.

1.5.3 Foundation Requirements

Foundations of structures shall be designed and constructed to support the structures during design flood conditions and shall provide the required support to prevent flotation, collapse, or permanent lateral movement under the load combinations specified in Section 1.6.2. Any part of the structure that provides structural support to that portion above the elevation specified by Table 2-1 or 4-1 shall meet applicable foundation requirements in this standard.

1.5.3.1 Geotechnical Considerations

Foundation design shall be based on the geotechnical characteristics of the soils and strata below the structure and on interactions between the soils and strata and the foundation. Foundation design shall account for instability and decreased structural capacity associated with soil consolidation, expansion, or movement; erosion and scour; liquefaction; and subsidence.

Geotechnical information necessary to complete the foundation design shall be obtained through geotechnical investigations of the site or from existing available data, such as investigations conducted at nearby project sites, regional studies conducted by government agencies, or other reliable sources.

1.5.3.2 Foundation Depth

The foundation shall extend to a depth based on geotechnical considerations to provide the support described in Sections 1.5.3 and 1.5.3.1, taking into account the erosion and local scour of the supporting soil based on an erosion analysis.

1.5.3.3 Foundation Walls

Foundation walls extending below the DFE shall be designed and constructed to account for

1. Hydrostatic, hydrodynamic, soil, wind, and other lateral loads acting during design flood conditions;
2. Buoyancy, dead load, live load, and other vertical loads acting during design flood conditions;
3. Strength, stability, and capacity of footings, other supporting structural elements, and underlying soils;
4. Strength, stability, and properties of materials used for foundation wall construction; and
5. Connections to footings and other supporting structural elements.

Masonry walls shall be designed and constructed in accordance with the requirements of ACI 530/

ASCE 5/TMS 402, *Building Code Requirements for Masonry Structures,* Ref. [2], and ACI 530.1/ASCE 6/TMS 602, *Specification for Masonry Structures,* Ref. [3].

Concrete shall be designed and constructed in accordance with the requirements of ACI 318, *Building Code Requirements for Structural Concrete,* Ref. [4].

1.5.3.4 Piers, Posts, Columns, or Piles

Piers, posts, columns, or piles used to elevate a structure above the DFE in flood hazard areas shall comply with all applicable foundation requirements of this standard. In Coastal High Hazard Areas and Coastal A Zones, piers, posts, columns, or piles used to elevate a structure shall be designed and constructed in accordance with Section 4. Connections between footings, mat, or raft foundations and piers, posts, and columns shall meet all applicable requirements of this standard.

1.5.4 Use of Fill

Fill shall be designed to be stable under conditions of flooding, including rapid rise and rapid drawdown of floodwaters, prolonged inundation, and flood-related erosion and scour. Use of fill in flood hazard areas other than High Risk Flood Hazard Areas shall be in accordance with Section 2.4. Use of fill in Coastal High Hazard Areas and Coastal A Zones shall be in accordance with Section 4.5.4.

1.5.5 Anchorage and Connections

The structure, including anchorage and connections, shall be designed to resist effects of vertical loads, including uplift and lateral loads in accordance with the load combinations specified in Section 1.6.2.

Stringers or beams shall be attached to the substructure or directly to piles, columns, piers, and walls with bolted or welded connections such that a continuous load path is maintained.

Washers shall be used under all nuts and bolt heads bearing directly on wood. All nuts, bolts, and washers shall be corrosion resistant. Notches at the tops of timber posts and piles shall not exceed 50% of the cross-section of the post or pile.

Adequate anchorage shall also be provided for storage tanks, sealed conduits and pipes, lined pits, sumps, and all other similar structures that have a negligible weight of their own or are subject to flotation or lateral movement during the design flood.

1.6 LOADS IN FLOOD HAZARD AREAS

1.6.1 General

Design of structures within flood hazard areas shall be governed by the loading provisions of ASCE 7 *Minimum Design Loads for Buildings and Other Structures,* Ref. [1].

Design and construction of structures located in flood hazard areas shall consider all flood-related loads and conditions, including the following: hydrostatic loads, hydrodynamic loads, wave action; debris impact; rapid rise and rapid drawdown of floodwaters; prolonged inundation; alluvial fan flooding; wave-induced and flood-related erosion and scour; deposition of sediments; ice flows and ice jams; and mudslides, in accordance with requirements of this standard if specified, or if not specified in this standard then in accordance with requirements approved by the authority having jurisdiction. Design considerations shall be documented and shall take into account the applicable flood-related loads and conditions, and concurrent loads, that will act on the structure.

1.6.2 Combination of Loads

Flood loads shall be combined with other loads as specified in the provisions for combinations of loads of ASCE 7 *Minimum Design Loads for Buildings and Other Structures,* Ref. [1] either by using the allowable stress design method load combinations or by using the strength design method load combinations.

2.0 BASIC REQUIREMENTS FOR FLOOD HAZARD AREAS THAT ARE NOT IDENTIFIED AS COASTAL HIGH HAZARD AREAS AND COASTAL A ZONES

2.1 SCOPE

The requirements of this section shall apply to new construction and substantial improvements in Flood Hazard Areas, including High Risk Flood Hazard Areas that are not identified as Coastal High Hazard Areas and Coastal A Zones.

2.2 DEVELOPMENT IN FLOODWAYS

Structures and fill shall not be constructed or placed in floodway areas unless it is demonstrated that those structures and fill will not, during the design flood, (1) increase the flood level, and (2) reduce the conveyance of the floodway.

If the design flood elevation has been determined and a floodway has not been designated, structures and fill shall not be constructed or placed unless it has been demonstrated that the cumulative effect of proposed structures and fill, combined with all other existing and anticipated development, will not increase the base flood elevation more than 1 ft.

2.3 ELEVATION REQUIREMENTS

Structures shall have the lowest floor (including basements) elevated to or above the Design Flood Elevation (DFE) in conformance with the requirements of Table 2-1. Enclosed areas used solely for parking, building access, or storage that comply with Section 2.6 are allowed below elevated buildings. Elevation requirements for other building components are found in Sections 5, 6, and 7.

TABLE 2-1. Minimum Elevation of the Top of Lowest Floor Relative to Base Flood Elevation (BFE) or Design Flood Elevation (DFE)—Flood Hazard Areas Other Than Coastal High Hazard Areas,[a] Coastal A Zones,[a] and High Risk Flood Hazard Areas[a]

Structure Category[b]	Minimum Elevation of Lowest Floor
I	DFE
II[c]	BFE + 1 ft or DFE, whichever is higher
III[c]	BFE + 1 ft or DFE, whichever is higher
IV[c]	BFE + 2 ft or DFE, whichever is higher

[a]Minimum elevations shown in Table 2-1 do not apply to Coastal High Hazard Areas and Coastal A Zones (see Table 4-1). Minimum elevations shown in Table 2-1 apply to other High Risk Flood Hazard Areas unless specific elevation requirements are given in Section 3 of this standard.

[b]See Table 1-1 for structure category descriptions.

[c]For nonresidential buildings and nonresidential portions of mixed-use buildings, the lowest floor shall be allowed below the minimum elevation if the structure meets the floodproofing requirements of Section 6.

Exception: Nonresidential structures with the lowest floor below the minimum elevation specified in Table 2-1 and nonresidential portions of mixed-use structures with the lowest floor below the minimum elevation specified by Table 2-1 shall be allowed in conformance with the floodproofing requirements of Section 6.

2.4 USE OF FILL

Structural fill, and nonstructural fill, shall not be placed in floodway areas unless in compliance with the requirements of Section 2.2.

2.4.1 Structural Fill

Structural fill shall not be used unless design and construction of the structural fill account for

1. Consolidation of the underlying soil under the weight of the fill and the structure;
2. Differential settlement due to variations in fill composition and characteristics; and
3. Slope stability and erosion control.

Fill used for structural support or protection shall be suitable for its intended use. Fill used to support or protect a structure shall be placed in lifts of not more than 12-in. loose thickness, with each lift compacted to at least 95% of its maximum Standard Proctor density (see Ref. [5]) or 90% of its maximum modified Proctor density (see Ref. [6]), unless a soils engineering report approved by the authority having jurisdiction specifies otherwise.

The side slopes of structural fill shall be no steeper than 1 on 1.5 (vertical/horizontal). Structural fill, including side slopes, shall be protected from scouring and erosion under flood conditions up to and including the design flood.

2.5 SLABS-ON-GRADE AND FOOTINGS

2.5.1 Use of Slabs-on-Grade

Use of slabs-on-grade is acceptable if the slab is installed on structural fill that is placed in conformance with Section 2.4 or is installed on undisturbed soil of adequate bearing capacity. The top of the slab shall be at or above the DFE as specified in Table 2-1. If turned down to act as footings, the bottom of the turned-down edges of the slab shall be installed at or below the depth of expected scour.

Reinforcement shall be sufficient to prevent breakup of the slab during design conditions, even if the soil under the slab is undermined by erosion. Slabs-on-grade installed on structural fill shall be placed so that there is no loss of supporting soil during the design flood conditions.

Exception: When located under an elevated building, slabs-on-grade shall not be reinforced or use turned-down edges.

2.5.2 Footing Design

Footings shall support the structure during design flood conditions, including prolonged inundation and scour and erosion if expected during design flood conditions, to prevent flotation, collapse, and lateral movement.

Footings that are also intended to act as grade beams shall comply with the provisions of Section 4.5.9.

2.6 ENCLOSURES BELOW THE DESIGN FLOOD ELEVATION

Enclosed areas that are used solely for parking, building access, or storage shall be permitted below the DFE provided the enclosed areas meet the requirements of this section.

2.6.1 Required Openings in Foundation Walls

Foundation walls that enclose an area below the DFE, and that do not meet the dry-floodproofing requirements of Section 6.2, shall contain openings to allow for automatic entry and exit of floodwaters during design flood conditions. These openings shall meet the requirements of Section 2.6.2.

2.6.1.1 Openings in Breakaway Walls

Openings to allow for the automatic entry and exit of floodwaters during design flood conditions shall be installed in breakaway walls in flood hazard areas other than Coastal High Hazard Areas. Openings shall meet the requirements of Section 2.6.2 or Section 4.6.2.

Openings in breakaway walls in Coastal High Hazard Areas shall not be required.

2.6.2 Design of Openings

Openings shall meet the nonengineered requirements of Section 2.6.2.1 or the engineered opening requirements of Section 2.6.2.2.

2.6.2.1 Nonengineered Openings

Nonengineered openings shall meet the following criteria:

1. There shall be a minimum of two openings on different sides of each enclosed area; if a structure has more than one enclosed area below the DFE, each area shall have openings;
2. The total net area of all openings shall be at least 1 square inch for each square foot of enclosed area;
3. The bottom of each opening shall be no more than 1 ft above the adjacent ground level;
4. Openings shall not be less than 3 in. in any direction in the plane of the wall;
5. Any louvers, screens, or other opening covers shall not block or impede the automatic flow of floodwaters into and out of the enclosed areas; and
6. Openings meeting requirements 1 through 5 above installed in doors and windows are acceptable; however, doors and windows are not deemed to meet the requirements of this section.

2.6.2.2 Engineered Openings

Engineered openings shall meet the following criteria:

1. Each individual opening, and any louvers, screens, or other covers, shall be designed to allow automatic entry and exit of floodwaters during design flood or lesser flood conditions;
2. There shall be a minimum of two openings on different sides of each enclosed area; if a structure has more than one enclosed area below the DFE, each area shall have openings;
3. Openings shall not be less than 3 in. in any direction in the plane of the wall;
4. The bottom of each required opening shall be no more than 1 ft above the adjacent ground level;
5. The difference between the exterior and interior floodwater levels shall not exceed 1 ft;
6. In the absence of reliable data on the rates of rise and fall, assume a minimum rate of rise and fall of 5 ft/h; where an analysis indicates the rates of rise and fall are greater than 5 ft/h, the total net area of the required openings shall be increased to account for the higher rates of rise and fall; where

an analysis indicates the rates of rise and fall are less than 5 ft/h, the total net area of the required openings shall remain the same or shall be decreased to account for the lower rates of rise and fall; and

7. The minimum total net area of the required openings in non-breakaway enclosure walls shall be calculated using the equation:

$$A_o = 0.033\,(1/c)(R)(A_e)$$

where

A_o = the total net area of openings required (in.^{2});
0.033 = coefficient (in.$^2 \cdot$ h/ft^3) corresponding to a factor of safety of 5.0;
c = opening coefficient of discharge given in Table 2-2;
R = worst case rate of rise and fall (ft/h); and
A_e = the total enclosed area (ft^2).

3.0 HIGH RISK FLOOD HAZARD AREAS

3.1 SCOPE

The requirements of Section 3 shall apply to new construction and substantial improvements in High Risk Flood Hazard Areas subject to one or more of the following hazards: alluvial fan flooding, flash floods, mudslides, erosion, high velocity flows, high velocity wave action, breaking wave heights greater or equal to 1.5 ft (Coastal High Hazard Area and Coastal A Zone) and damage-causing ice or debris. In addition to the requirements of Section 3, the basic requirements of Section 2 shall apply to High Risk Flood Hazard Areas other than Coastal High Hazard Areas and Coastal A Zones.

Table 2-2. Flood Opening Coefficient of Discharge

Opening Shape and Condition	c
Circular, unobstructed during design flood	0.60
Rectangular, long axis horizontal, short axis vertical, unobstructed during design flood	0.40[a]
Square, unobstructed during design flood	0.35
Rectangular, short axis horizontal, long axis vertical, unobstructed during design flood	0.25[b]
Other shapes, unobstructed during design flood	0.30

[a]When the horizontal dimension is twice or more the vertical dimension, use 0.4; as the dimensions approach a square, interpolate from 0.4 to 0.35.

[b]When the horizontal dimension is half or less the vertical dimension, use 0.25; as the dimensions approach a square, interpolate from 0.25 to 0.35.

3.2 ALLUVIAL FAN AREAS

Structures shall not be constructed at the apex of an alluvial fan and in the fan's meandering flow paths. Construction in other areas of the alluvial fan shall meet the following requirements:

1. The elevation of the lowest floor shall be a minimum of 1 ft above the highest adjacent grade, or higher, if required on a community's flood hazard map;
2. Foundations shall be designed and constructed to resist scour caused by the actual flow velocity but not less than 5 ft/sec. Determination of actual flow velocities shall be based on a review of a community's flood hazard map and flood hazard study or on hydraulic calculations; and
3. Design and construction shall resist all load combinations specified in Section 1.6.2.

3.2.1 Protective Works in Alluvial Fan Areas

Structures shall not be allowed in alluvial fan areas unless protective works (whole alluvial fan flood damage reduction project) have been designed and constructed:

1. To safely pass the design flood at the apex, within the capacity of the constructed channel(s);
2. Such that it does not divert flood flows and debris toward other structures, nor increase flood velocities and depths elsewhere on the alluvial fan; and
3. Such construction satisfies the requirements of Section 1.4.2, and a maintenance and operations plan for the protective works is provided.

3.3 FLASH FLOOD AREAS

Structures shall not be constructed in areas subject to flash flooding equal to or less than design flood conditions.

Areas suspected of being subject to flash floods shall be investigated to obtain historical information on past events. The investigation shall also include analysis of historic rainfall and runoff data for the watershed. Results of such analyses shall be documented in an engineering report, which defines the methodology and data used to conclude whether the area in question has the potential for flash flooding.

3.3.1 Protective Works in Flash Flood Areas

Structures shall not be constructed in areas subject to flash floods, unless protective works have been determined to provide protection during the design flood event, where such construction satisfies the requirements of Section 1.4.2, and where a maintenance and operations plan for the protective works has been provided.

3.4 MUDSLIDE AREAS

Structures shall not be constructed in areas subject to mudslides during periods of rainfall and runoff. Areas suspected of being subject to inundation by mudslides shall be investigated to obtain historical information on past flood events. The investigation shall also include analysis of the source area for potential overland or channel erosion, bank failure, hillslope failure, and rainfall/runoff potential. Results of such analyses shall be documented in an engineering report, which defines the methodology and data used to conclude whether the area in question has potential for future mudslides.

3.4.1 Protective Works in Mudslide Areas

Structures shall not be constructed in areas subject to mudslides, unless protective works have been determined to provide protection during the design flood event, such construction satisfies the requirements of Section 1.4.2, and a maintenance and operations plan for the protective works has been provided.

3.5 EROSION-PRONE AREAS

Structures shall not be constructed within flood hazard areas subject to erosion from such phenomena as caving banks, meandering streams, or eroding shorelines, where such erosion is predicted to affect the structure unless the structure is protected as specified in Section 3.5.1.

Erosion-prone areas shall be determined by analyzing available studies, historical data, watershed trends, average annual erosion rates, wave effects, flood velocities and duration of flow, geotechnical data, and existing protective works. Results of these analyses shall be documented in an engineering report, which defines the data and methodology used to identify erosion-prone areas.

3.5.1 Protective Works in Erosion-Prone Areas

The limits of an erosion-prone area shall be subject to revision where protective works have been designed and constructed to control erosion processes during all flow and wave conditions up to and including the design flood, and where a maintenance and operations plan for the protective works has been provided.

3.6 HIGH VELOCITY FLOW AREAS

High velocity flow areas shall be identified from a community's flood hazard map or flood hazard study or from hydraulic analyses. The results of such analyses shall be documented in an engineering report, which defines the methodology and data used to conclude whether a site is susceptible to high velocity flows.

3.6.1 Protective Works in High Velocity Flow Areas

Structures shall not be constructed in high velocity flow areas, unless protective works have been determined to provide protection during the design flood event, such construction satisfies the requirements of Section 1.4.2, and a maintenance and operations plan for the protective works has been provided.

3.7 AREAS SUBJECT TO WAVE ACTION

3.7.1 Coastal High Hazard Areas and Coastal A Zones

Structures shall not be constructed in Coastal High Hazard Areas and Coastal A Zones unless the design and construction meet the requirements of Section 4.

3.7.2 Other High Velocity Wave Action Areas

Other High Velocity Wave Action Areas include noncoastal flood hazard areas subject to wind-driven waves greater than or equal to 3 ft in height, including large lakes and expansive riverine flood hazard areas. Structures shall not be constructed in other areas susceptible to high velocity wave action, unless the design and construction meet the requirements of Section 4.

3.8 ICE JAM AND DEBRIS AREAS

Structures shall not be constructed within flood hazard areas that are subject to transportation of damage-causing ice or debris during floods up to and including the design flood.

The potential for ice or debris capable of inducing or causing loads exceeding design loads shall be identified from a community's flood hazard map or flood hazard study or from hydraulic and other analyses. The results of such analyses shall be documented in an engineering report, which defines the methodology and data used to conclude whether a site is susceptible to ice jams and debris effects.

3.8.1 Protective Works in Ice Jam and Debris Areas

Structures in ice jam and debris areas shall have protective works to provide protection during the design flood event and meet the requirements of Section 1.4.2. The maintenance and operations plan for the protective works shall be provided.

4.0 COASTAL HIGH HAZARD AREAS AND COASTAL A ZONES

4.1 SCOPE

The requirements of Section 4 shall apply to new construction and substantial improvements in Coastal High Hazard Areas and Coastal A Zones

4.1.1 Identification of Coastal High Hazard Areas and Coastal A Zones

For the purposes of this standard, "Coastal High Hazard Areas" shall mean those locations where an area has been designated as subject to high velocity wave action on a community's flood hazard map (V Zones), or

1. Where the stillwater depth of the base flood above the eroded ground elevation is greater than or equal to 3.8 ft, i.e., sufficient to support a wave height equal to or greater than 3 ft and where conditions are conducive to the formation and propagation of such waves; or
2. Where the eroded ground elevation under base flood conditions is 3 ft or more below the maximum wave runup elevation.

For the purposes of this standard, "Coastal A Zones" shall mean those locations where the stillwater depth of the base flood above the eroded ground elevation is greater than or equal to 1.9 ft (sufficient to support a wave height equal to or greater than 1.5 ft), and where conditions are conducive to the formation and propagation of such waves.

4.2 GENERAL

Designs for Coastal High Hazard Areas and Coastal A Zones shall account for the following:

1. Waves breaking against the side or underside of the structure;
2. Drag, inertia, and other wave-induced forces acting on structural members supporting elevated structures;
3. Uplift forces from breaking waves striking the undersides of structures;
4. Wave runup forces including those deflected by the structure; and
5. Erosion and scour.

4.3 SITING

New construction and substantial improvements within Coastal High Hazard Areas and Coastal A Zones shall

1. Be located landward of the reach of mean high tide;
2. Be sited landward of shoreline construction setbacks; and
3. Not remove or otherwise alter sand dunes and mangrove stands, unless the alterations will not reduce the wave and flow dissipation characteristics of the sand dunes or mangrove stands.

4.4 ELEVATION REQUIREMENTS

The bottom of the lowest horizontal structural member of the lowest floor shall be at or above the Design Flood Elevation (DFE), in conformance with the requirements of Table 4-1. The actual required height above the DFE shall be determined by the structure category and the orientation of the lowest horizontal structural member relative to the direction of wave approach. Piles, pile caps, footings, mat or raft foundations, grade beams, columns, and shear walls designed and constructed in accordance with Section 4.5 shall not be required to meet the elevation requirements of Table 4-1.

4.5 FOUNDATION REQUIREMENTS

4.5.1 General

Foundation systems located in Coastal High Hazard Areas and Coastal A Zones shall be designed to minimize forces acting on that system. Foundation

TABLE 4-1. Minimum Elevation of Bottom of Lowest Supporting Horizontal Structural Member of Lowest Floor Relative to Base Flood Elevation (BFE) or Design Flood Elevation (DFE)—Coastal High Hazard Areas and Coastal A Zones

Structure Category[a]	Member Orientation Relative to the Direction of Wave Approach	
	Parallel[b]	Perpendicular[b]
I	DFE	DFE
II	DFE	BFE + 1 ft or DFE, whichever is higher
III	BFE + 1 ft or DFE, whichever is higher	BFE + 2 ft or DFE, whichever is higher
IV	BFE + 1 ft or DFE, whichever is higher	BFE + 2 ft or DFE, whichever is higher

[a] See Table 1-1 for structure category descriptions.
[b] Orientation of lowest horizontal structural member relative to the general direction of wave approach: parallel shall mean less than or equal to +20 degrees from the direction of approach; perpendicular shall mean greater than +20 degrees from the direction of approach.

systems shall be free of obstructions and attachments that will transfer flood forces to the structural system or that will restrict or eliminate free passage of high velocity flood waters and waves during design flood conditions.

Structures shall be supported on piles, columns, or walls serving as shear walls. Spread footing, mat, or raft foundations shall not be used unless the top of the spread footing, mat, or raft foundation is below the eroded ground elevation. Piles shall extend upward to a point at or above the DFE, as required by Table 4-1. Columns shall be connected to and extend upward from the spread footing, mat, or raft foundation to a point at or above the DFE, as required by Table 4-1. Shear walls shall comply with the requirements of Section 4.5.11.

Where surface or subsurface conditions consist of nonerodible soil that prevents the use of pile or deeply embedded column foundations, spread footing or mat foundations shall be permitted provided they are anchored, if necessary to prevent sliding, uplift, or overturning, to nonerodible soil with sufficient strength to withstand forces from the combination of loads in Section 1.6.2.

4.5.2 Special Geotechnical Considerations

In addition to the requirements of Section 1.5.3, foundation design shall account for instability and decreased structural capacity associated with erosion due to wind, waves, currents, local scour, storm-induced erosion, and shoreline movement.

4.5.3 Foundation Depth

The foundation shall extend to a depth sufficient to provide the support required in Section 1.5.3, taking into account the erosion and scour of the supporting soil during the design flood, and shoreline movement, as predicted by an erosion analysis.

4.5.4 Use of Fill

Fill material used for structural support shall not be permitted in Coastal High Hazard Areas and Coastal A Zones. Placement of nonstructural fill for minimal site grading and landscaping, and to meet local drainage requirements, shall be permitted. Placement of nonstructural fill under and around a structure for dune construction or reconstruction shall be permitted if the fill will not result in wave runup, ramping, or deflection of floodwaters that cause damage to structures.

4.5.5 Pile Foundations

Except as provided for under Section 4.5.1, all foundations constructed in erodible soils shall be founded on piles. Piles that are jetted or installed in an augured excavation shall be seated by driving.

In erodible soils, pile tip penetration shall be to a minimum depth of 10 ft below mean water level (−10 ft MWL), unless the design demonstrates that pile penetration to a shallower depth will provide the support and stability required by Section 4.5.3. In the event that unexpected conditions are encountered during construction and refusal or design friction capacity is not reached during pile installation, additional geotechnical investigations and a revised pile design shall be completed.

The design shall consider that local scour and liquefaction of the erodible soil during design flood conditions will render as nonsupportive, the soil at least to a depth of two times the pile diameter (round pile) or two times the diagonal (rectangular pile) below the ground elevation at the point of pile penetration after flood- or storm-induced erosion has been considered. This nonsupportive soil shall not be considered in the design. Calculation of local scour effects during design flood conditions shall include the interactive effects of pilings or other foundation elements in close proximity to one another.

4.5.5.1 Attachments to Piles

Pile design shall account for the additional loads resulting from any attachments to the piles and shall account for increased scour around the base of the pile.

Bracing used for lateral support of pile foundation systems shall be in accordance with Section 4.5.10.

4.5.5.2 Piles Terminating in Caps at or Below Grade

Foundations composed of a number of single piles or pile clusters terminating in reinforced concrete pile caps at or below grade shall be designed for the combination of loads in Section 1.6.2. Embedment of the pile into the pile cap shall be sufficient to resist separation of the two. Embedment shall not be less than that specified by Section 4.5.6.

The pile cap shall be designed and constructed to be structurally stable, without relying on supporting soil around or under the cap. Pile-to-pile cap and pile cap-to-column connections shall be designed to withstand expected hydrodynamic forces including wave and debris impact.

4.5.5.3 Piles Extending to Superstructure (Structure Framing)

The portion of a pile that extends above the eroded ground elevation to elevate a structure shall be designed as a column. Pile spacing shall take into account the design bearing capacity, uplift, and overturning resistance.

Bracing shall be provided in accordance with Section 4.5.10.

4.5.5.4 Wood Piles

Wood piles shall be preservative treated in accordance with Section 5. Consideration shall be given to the use of pile fittings at the butt, tip, and at designated intervals along the pile length for the protection of piles during installation. Round timber piles shall conform to ASTM D 25 *Standard Specification for Round Timber Piles,* Ref. [7].

Round piles shall have a minimum diameter of 8 in. at the tip. The minimum size for square piles shall be 8 in. on a side where the exposed pile length (after accounting for scour and erosion) is equal to or less than 12 ft and shall be 10 in. on a side where the exposed pile length (after accounting for scour and erosion) is greater than 12 ft.

Wood piles that are directly connected to beams supporting an elevated lowest floor shall be individually secured thereto by means of at least two hot-dipped galvanized or stainless steel bolts, nuts, and washers, of sufficient size and number to resist the forces resulting from the combinations of loads in Section 1.6.2.

4.5.5.5 Steel H Piles

Steel piles of rolled HP sections or built-up sections shall consist of a corrosion resistant material or be protected from corrosion by a corrosion resistant coating or by cathodic protection, in accordance with Section 5. Built-up sections shall have a web thickness equal to the flange thickness; the web and flange shall be continuously welded together.

The metal thickness shall be based on the loss of section due to corrosion, unless corrosion protection is provided in the form of concrete, bituminous, or plastic (epoxy) coatings or cathodic protection. The minimum allowable thickness of the metal shall be 0.4 in. Damage to coatings during installation shall be avoided, and, if damage occurs, repairs shall be made in accordance with manufacturers' recommendations and applicable codes, standards, and regulations.

Pile tip reinforcing, splicing, fittings, and cap plates shall be provided, as required.

4.5.5.6 Concrete-Filled Steel Pipe Piles and Shells

Steel components of concrete-filled steel pipe piles and shells shall be protected with a corrosion resistant coating, in accordance with Section 5.

Pipe for concrete-filled steel pipe piles shall conform to ASTM A 252 *Standard Specification for Welded and Seamless Steel Pipe Piles,* Ref. [8]. Pile tip reinforcing, splicing, fittings, and cap plates shall be specified as required. Flat steel plates closing the tip of pipe piles shall be of a diameter not greater than 3/4 in. more than the outside diameter of the pipe.

The slump of concrete and maximum size of coarse aggregate used in concrete to be placed in piles shall be specified with consideration given to the difficulty of placement conditions.

4.5.5.7 Prestressed Concrete Piles and Precast Concrete Piles

Prestressed concrete piles and precast concrete piles shall be designed, manufactured, and installed in accordance with Section 5.

Pile dimensions shall be 10 in. minimum for piles of uniform section and 8 in. minimum for tapered piles.

For piles subject to exposure from brackish water, seawater, or spray from these sources, cover for reinforcement shall be not less than 3 in. for precast concrete piles and not less than 2-1/2 in. for prestressed concrete piles.

4.5.5.8 Cast-in-Place Concrete Piles

Concrete and steel reinforcement used in cast-in-place concrete piles shall conform to the requirements of ACI 318 *Building Code Requirements for Structural Concrete,* Ref. [4].

4.5.6 Pile Design

4.5.6.1 Pile Capacity

Piles shall be designed to carry the loads imposed by the combinations of loads in Section 1.6.2 and to withstand installation forces.

Unless exceeded by moments indicated by analysis of applied loads, the piles shall be designed for a minimum moment produced by an eccentricity of 0.10 times the equivalent diameter of the pile times the axial load on the pile.

The minimum required lateral resistance of an individual pile shall be at least 5% of the axial load on the pile.

4.5.6.2 Capacity of the Supporting Soils

Soil values pertaining to friction, end bearing resistance, and settlement of single piles and pile groups shall be based on the geotechnical characteristics of the soil as required by Section 1.5.3.1.

For piles spaced more than three pile diameters apart, measured center to center, the diameter of the soil that shall be assumed reacting laterally on each pile shall have a maximum equivalent diameter equal to three times the diameter of the pile.

4.5.6.3 Minimum Penetration

Pile penetration into acceptable bearing strata shall be a minimum depth sufficient to allow distribution of the pile load to the supporting soils, including a consideration for reduction in soils due to the effects of scour and erosion, in accordance with Section 1.5.3.2 and/or Section 4.5.3.

4.5.6.4 Foundation Pile Spacing

Pile spacing shall be not less than 8 ft as measured center to center, unless otherwise justified by a geotechnical analysis and the foundation design.

4.5.6.5 Pile Caps

Pile caps shall be constructed with their tops at or below grade. The design shall provide for the effects of scour and erosion. Piles shall be designed to carry the total superimposed vertical load from the structure above with no allowance made for the supporting value of the soil under the pile caps. Pile caps and piles shall be designed to resist lateral flood loads acting on pile caps and pile sections exposed by erosion and scour.

4.5.6.6 Timber Pile Connections

For timber piles, bolts for cap-to-pile or beam-to-pile connections shall be 5/8 in. in diameter minimum. Bolt holes shall be staggered with a maximum diameter of 1/16 in. greater than the bolt diameter. The dimension from the edge of the holes to the pile or beam edge shall be 2 in. minimum. Notching of pile tops shall not exceed 50% of the pile cross section. Other pile-to-beam connections are acceptable provided they are demonstrably equal to or superior than the applications and do not depend upon nailing for attachment of individual members.

For timber piles not in tension, and connected to timber caps, the tops of the piles shall be secured to the caps with spiral-drive drift bolts, metal plates, or bolted timber scabs.

For timber piles not in tension, and connected to concrete caps, the tops of the piles shall have a minimum 4 in. embedment into the concrete pile caps.

For timber piles in tension, piles shall be embedded into pile caps a minimum of 6 in., with a connection designed for tension made to pile caps. Connections to timber caps shall be made with timber scabs or metal straps, and headed bolts.

Connections of timber piles in tension to concrete pile caps shall have the tops of the piles embedded to satisfy requirements of shear stress in the timber (parallel to grain) and shear stress in the concrete. Connections shall be made with metal straps, headed bolts, or other forms of positive tension resisting devices that develop the necessary shear in the concrete without causing failure of the wood.

4.5.6.7 Steel Pile Connections

For steel piles not in tension, and where structural design of piles depends on bending in the piles for stability, the tops of steel piles shall be tied into concrete caps with reinforcing steel or structural sections welded to the pile, and lapped with the cap reinforcement. The minimum lap shall be 6 in.

For steel piles in tension, the tops of steel piles shall be tied into concrete caps with reinforcing steel or structural sections welded to the pile, and lapped with the cap reinforcement. The minimum lap shall be 6 in. Bond stress between concrete and steel shall not exceed $0.02 f'_c$, where f'_c is the specified compressive strength of concrete.

4.5.6.8 Concrete Pile Connections

For concrete piles not in tension, the tops of the piles shall have a minimum 3 in. embedment into the pile cap.

For concrete piles in tension, the tops of concrete piles in concrete caps shall be adequately doweled and embedded into the cap to resist tension loads. The tops of piles shall have a minimum 3 in. embedment into the pile cap.

4.5.6.9 Pile Splicing

Splices shall be constructed to provide and maintain the alignment and position of the component parts of the pile during installation and subsequent thereto. Splices shall be of adequate strength to transmit the axial and lateral loads, and the moments at the section involved.

Pile splices that cannot be visually inspected after pile installation shall develop the greater of at least 50% of the capacity of the pile or the moment, shear, and tension that would result from an assumed eccentricity of the resultant pile load of 3 in.

In areas subject to high-velocity wave action and subject to flood-related erosion, timber pile splicing shall be made below the scour elevation, as determined by an erosion analysis.

4.5.6.10 Mixed Types of Piling and Multiple Types of Installation Methodology

The use of mixed types or capacities of piling and different types of installation equipment or methods shall consider, in addition to the relative lateral load capacities, an analysis of the additional effects on the superstructure of differential elastic shortening and settlement.

4.5.7 Posts, Piers, and Columns

Columns, including wood posts, reinforced masonry columns and piers, and reinforced concrete columns and piers, as well as associated connections, shall be designed and constructed to resist wind, water, wave, erosion, and other flood-related forces, in accordance with Section 1.6. Column spacing shall take into account the design bearing capacity, uplift, and overturning resistance but shall be spaced not less than 8 ft center to center. Where founded on erodible soils, supports shall extend to at least −10 ft MWL (10 ft below mean water level) or shall be supported by and anchored to a pile, spread footing, mat, or raft foundation meeting the requirements of Section 4.5.1.

4.5.7.1 Wood Posts

Wood posts shall be 10 in. by 10 in. minimum for square posts or 10 in. in diameter for round posts. Wood posts shall be preservative treated in conformance with Section 5. Where founded on nonerodible soil, posts shall be adequately embedded and anchored to the footing, which shall be anchored to the substrata to prevent pullout during design flood conditions.

4.5.7.2 Reinforced Masonry Columns

Reinforced masonry columns shall comply with ACI 530/ASCE 5/TMS 402, *Building Code Requirements for Masonry Structures,* Ref. [2], and ACI 530.1/ASCE 6/TMS 602, *Specification for Masonry Structures,* Ref. [3].

4.5.7.3 Reinforced Concrete Columns

Concrete columns shall comply with ACI 318, *Building Code Requirements for Structural Concrete,* Ref. [4].

4.5.8 Footings, Mats, Rafts, and Slabs-on-Grade

Where permitted in accordance with Sections 2 and 4, spread footing, mat, raft, and other concrete slab foundation elements at or below grade shall be reinforced and supported directly on the soil or on grade beams that are in turn supported on the soil. The design shall provide for the effects of scour and erosion, and additional lateral and vertical loads from flood forces transferred to the structure through these elements.

4.5.9 Grade Beams

Grade beams shall be constructed with their upper surface at or below natural grade, shall be structurally connected to the primary foundation system to provide additional lateral support, and shall be structurally independent of decks, patios, and concrete pads. Grade beams shall be designed to perform their structural function without the aid of supporting earth and during exposure to flood forces.

Design of grade beams and other foundation elements shall provide for the effects of lateral bracing offered by grade beams on the structure.

Grade beams shall be firmly attached to vertical members to transfer all vertical and lateral forces acting on the grade beam as a result of flooding, including wave action and the effects of debris, scour, and erosion. The design of the grade beam members shall include biaxial beam action to support the vertical and lateral forces acting the full length of members in the event they are undermined by scour and erosion.

4.5.10 Bracing

Unless the design determines bracing is not required, bracing shall be provided where more than 8 ft of the pile or column is exposed above the eroded ground elevation. Bracing and its connections to the primary vertical structural members shall be designed to withstand the lateral force of waves and debris impact. The braces shall be designed to resist forces acting both in the plane of brace resistance and perpendicular to the plane of brace resistance. The foundation system shall be designed to account for the additional loads transferred from the bracing to the main supporting structures. Cross bracing perpendicular to the primary direction of wave and hydrodynamic forces shall be restricted to tensile bracing using steel rods or steel cables. Cross bracing parallel to the primary direction of wave and hydrodynamic forces shall not be restricted.

Steel rods used for cross bracing shall be galvanized or of noncorrosive material with a minimum diameter of 1/2 in. An equivalent system of corrosion-resistant high tensile steel cables and turnbuckles may be used in lieu of solid rods. Where rods or cables are used for bracing, they shall be tied together with a clamp at the center cross-point.

The smallest nominal dimension of any wood member used for compression cross bracing shall be not less than 4 in. The smallest nominal dimension of any wood member used for tension-only cross bracing shall be 2 in.

Cross bracing shall be attached to the main vertical structural elements with bolts, nuts, and plate washers. Connections to the main vertical structural elements shall be within 12 in. of the lowest elevated floor support beams and as near to grade as possible.

Knee braces are permitted in all directions relative to flood forces and shall extend not more than 3 ft below the DFE.

4.5.11 Shear Walls

Shear walls below the DFE shall be oriented parallel to the direction of wave approach where possible and shall be staggered so as not to form a continuous shear wall or an enclosed area. For the purpose of this provision, parallel shall mean less than or equal to ±20 degrees from the direction of wave approach. If shear walls are oriented other than parallel to the direction of wave approach, an unobstructed area equal to one-half the area blocked by the shear wall shall be provided adjacent to each shear wall.

4.6 ENCLOSED AREAS BELOW DESIGN FLOOD ELEVATION

Enclosed areas below the DFE shall be allowed only where all of the following conditions are met:

1. Enclosures shall be designed and constructed as breakaway construction that will not adversely affect any structure by redirecting floodwaters or by producing debris capable of causing damage to structures; and
2. Enclosed areas shall be used solely for parking, building access, or storage

4.6.1 Breakaway Walls

Breakaway walls and other similar nonload bearing components, including open-wood lattice work and insect screening, shall be designed and constructed to fail under base flood or lesser conditions, without imparting additional flood loads to the foundation or superstructure and without producing debris damage to the structure or adjacent structures. Breakaway walls and their connections shall be designed in accordance with the requirements of Section 5.3.2.3 of ASCE 7 *Minimum Design Loads for Buildings and Other Structures,* Ref. [1]. Utilities and attendant equipment shall not be mounted on, pass through, or be located along breakaway walls.

4.6.2 Openings in Breakaway Walls in Coastal A Zones

Openings for the entry and exit of floodwaters shall not be required in breakaway walls located in Coastal High Hazard Areas.

Openings to allow for the automatic entry and exit of floodwaters during design flood conditions shall be installed in breakaway walls in Coastal A Zones. The minimum total net area of the openings shall conform with Section 2.6.2.1 or shall be calculated using the equation

$$A_o = 0.007\ (1/c)(R)(A_e)$$

where

A_o = the total net area of openings required (in.^{2});
0.007 = coefficient (in.^{2} · h/ft^3) corresponding to a factor of safety of 1.0;
c = opening coefficient of discharge given in Table 2-2;
R = worst case rate of rise and fall (ft/hr); and
A_e = the total enclosed area (ft^2).

4.7 EROSION CONTROL STRUCTURES

Bulkheads, seawalls, revetments, and other erosion control structures shall not be connected to the foundation or superstructure of a structure. Bulkheads, seawalls, revetments, and other erosion control structures shall be designed and constructed so as to not focus or increase flood forces or erosion impacts on the foundation or superstructure of any structure.

4.8 DECKS, CONCRETE PADS, AND PATIOS

Decks, concrete pads, and patios below the DFE shall be permitted beneath or adjacent to a structure, provided they are structurally independent of the primary structural foundation system of the structure and provided they do not adversely affect adjacent structures through redirection of floodwaters or debris.

Decks, concrete pads, and patios shall not transfer flood loads to the main structure, shall be constructed to break away cleanly during design flood conditions, shall be friable, and shall not produce debris capable of causing significant damage to any structure.

Reinforcing, including welded wire fabric, shall not be used in order to minimize the potential for concrete pads being a source of debris. Pad thickness shall not exceed 4 in.

Decks and patios subject to floating shall be adequately anchored.

5.0 MATERIALS

5.1 GENERAL

New construction and substantial improvements in flood hazard areas shall be constructed with flood-damage–resistant materials below the elevations specified in Table 5-1. Flood-damage–resistant materials shall have sufficient strength, rigidity, and durability to adequately resist all flood-related and other loads, unless designed to break away or as permitted elsewhere in this standard.

Exposed structural and nonstructural construction materials, including connections, shall be capable of resisting damage, deterioration, corrosion, or decay due to precipitation, wind-driven water, salt spray, or other corrosive agents known to be present.

Table 5-1. Minimum Elevation, Relative to Base Flood Elevation (BFE) or Design Flood Elevation (DFE), Below which Flood-Damage–Resistant Materials Shall Be Used

		Coastal High Hazard Areas and Coastal A Zones	
Structure Category[a]	Flood Hazard Areas	Orientation Parallel[b]	Orientation Perpendicular[b]
I	DFE	DFE	DFE
II	BFE + 1 ft or DFE, whichever is higher	BFE + 1 ft or DFE, whichever is higher	BFE + 2 ft or DFE, whichever is higher
III	BFE + 1 ft or DFE, whichever is higher	BFE + 2 ft or DFE, whichever is higher	BFE + 3 ft or DFE, whichever is higher
IV	BFE + 2 ft or DFE, whichever is higher	BFE + 2 ft or DFE, whichever is higher	BFE + 3 ft or DFE, whichever is higher

[a] See Table 1-1 for structure category descriptions.
[b] Orientation of lowest horizontal structural member relative to the general direction of wave approach: parallel shall mean less than or equal to +20 degrees from the direction of approach; perpendicular shall mean greater than +20 degrees from the direction of approach.

Structural and nonstructural construction materials, including connections, below the elevations specified in Table 5-1 shall be capable of resisting damage, deterioration, corrosion, or decay due to direct and prolonged contact with floodwaters, associated with design flood conditions.

Materials used in new construction in flood hazard areas shall have sufficient strength, rigidity, and durability to adequately resist all flood-related and other loads during installation.

5.2 SPECIFIC MATERIALS REQUIREMENTS FOR FLOOD HAZARD AREAS

5.2.1 Metal Connectors and Fasteners

Metal plates, connectors, screws, bolts, nails, and other fasteners exposed to direct contact by flood water, precipitation, or wind-driven water shall be stainless steel or equivalent corrosion-resistant material, or hot-dip galvanized in accordance with ASTM A 123/A 123M *Standard Specification for Zinc (Hot-Dip Galvanized) Coatings on Iron and Steel Products,* Ref. [9], ASTM A 153/A 153M *Standard Specification for Zinc Coating (Hot-Dip) on Iron and Steel Hardware,* Ref. [10], ASTM A 653/A 653M *Standard Specification for Steel Sheet, Zinc-Coated (Galvanized) or Zinc-Iron Alloy-Coated (Galvannealed) by the Hot-Dip Process,* Ref. [11], or ASTM A 924/A 924M *Standard Specification for General Requirements for Steel Sheet, Metallic-Coated by the Hot-Dip Process,* Ref. [12].

5.2.2 Structural Steel

Steel piles shall conform to ASTM A 36/A 36M *Standard Specification for Carbon Structural Steel,* Ref. [13], ASTM A 572/A 572M *Standard Specification for High-Strength Low-Alloy Columbium-Vanadium Structural Steel,* Ref. [14], or ASTM A 690/A 690M *Standard Specification for High-Strength Low-Alloy Steel H-Piles and Sheet Piling for Use in Marine Environments,* Ref. [15].

5.2.2.1 Corrosive Environments

Structural steel exposed to direct contact with salt water, salt spray, or other corrosive agents known to be present shall be hot-dipped galvanized after fabrication. Secondary components such as angles, bars, straps, and anchoring devices shall be stainless steel or hot-dipped galvanized after fabrication, in accordance with Section 5.2.1.

5.2.2.2 Noncorrosive Environments

In areas where salt spray and other corrosive agents are known not to be present, exposed structural steel either shall meet the requirements of Section 5.2.2.1 or shall be primed, coated, plated, or otherwise protected against corrosion due to direct contact with floodwaters, precipitation, or wind-driven water.

Secondary components such as angles, bars, straps, and anchoring devices shall be stainless steel or hot-dipped galvanized after fabrication, in accordance with Section 5.2.1.

Damage to protective finishes and coatings caused by handling or installation shall be repaired using procedures that result in protection equivalent to the requirements stated above.

5.2.3 Concrete

Ingredients of concrete, including admixtures and reinforcing steel; quality of concrete; and the design and construction thereof shall comply with ACI 318 *Building Code Requirements for Structural Concrete,* Ref. [4], with special consideration for requirements concerning durability, including protection from chlorides and sulfates found in a saltwater environment.

5.2.4 Masonry

Materials used in masonry construction, including masonry units, mortar, grout, reinforcing steel and accessories; quality of masonry; and the design and construction thereof shall comply with ACI 530/ASCE 5/TMS 402, *Building Code Requirements for Masonry Structures,* Ref. [2], and ACI 530.1/ASCE 6/TMS 602 *Specification for Masonry Structures,* Ref. [3].

5.2.5 Wood and Timber

Wood and timber members, exposed or enclosed, solid or built-up, shall be naturally decay resistant or pressure treated with preservatives to resist damage, deterioration, or decay due to insect infestation, decay-producing fungi, or, when exposed, marine borers; or direct and prolonged wetting from floodwaters, precipitation, wind driven water, moisture, salt spray, or other corrosive agents known to be present. The design and construction of wood and timber shall comply with *National Design Specification for Wood Construction,* Ref. [16].

5.2.6 Finishes

Interior finishes and trim shall be flood-damage–resistant material.

6.0 DRY AND WET FLOODPROOFING

6.1 SCOPE

This section addresses design and construction requirements for floodproofing new construction and substantial improvements in flood hazard areas.

Design of floodproofing measures shall take into consideration and account for the flood loads and combination of loads in Section 1.6, and for nature of flood-related hazards; frequency, depth, and duration of flooding; rate of floodwater rise and fall; floodwater temperature; soil characteristics; flood-borne contaminants and debris; flood warning time; access to and from floodproofed areas; structure occupancy and use; and functional dependence.

6.2 DRY FLOODPROOFING

Dry floodproofing shall be accomplished through the use of flood-damage-resistant materials and techniques that render the dry-floodproofed portions of a structure substantially impermeable to the passage of floodwater below the elevations specified in Table 6-1. Sump pumps shall be provided to remove water accumulated due to any passage of vapor and seepage of water during the flooding event. Sump pumps shall not be relied upon as a means of dry floodproofing. All materials below the elevations specified in Table 6-1 shall conform with the requirements of Section 5.

TABLE 6-1. Minimum Elevation of Floodproofing, Relative to Base Flood Elevation (BFE) or Design Flood Elevation (DFE)—Outside of High Risk Flood Hazard Areas

Structure Category[a]	Minimum Elevation of Floodproofing[b]
I	BFE + 1 ft or DFE, whichever is higher
II[c]	BFE + 1 ft or DFE, whichever is higher
III	BFE + 1 ft or DFE, whichever is higher
IV	BFE + 2 ft or DFE, whichever is higher

[a] See Table 1-1 for structure category descriptions.
[b] Wet or dry floodproofing shall extend to the same level.
[c] Dry floodproofing of residential buildings and residential portions of mixed-use buildings shall not be permitted.

6.2.1 Dry Floodproofing Limitations

Dry floodproofing of nonresidential structures and nonresidential areas of mixed-use structures shall not be allowed unless such structures are located outside of High Risk Flood Hazard Areas, Coastal High Hazard Areas, and Coastal A Zones. Dry floodproofing of residential structures or residential areas of mixed-use structures shall not be permitted.

Dry floodproofing shall be limited to the following:

1. Where flood velocities adjacent to the structure are less than or equal to 5 ft/sec during the design flood; and
2. If human intervention is proposed, where conformance with the limitations of Section 6.2.3 is provided.

6.2.2 Dry Floodproofing Requirements

Dry floodproofed areas of structures shall

1. Be designed and constructed so that any area below the applicable elevation specified in Table 6-1, together with attendant utilities and sanitary facilities, is flood resistant with walls that are substantially impermeable to the passage of water. Walls, floors, and flood shields shall be designed and constructed to resist hydrostatic, hydrodynamic, and other flood-related loads, including the effects of buoyancy resulting from flooding to the elevation listed in Table 6-1;
2. Have any soil or fill adjacent to the structure compacted and protected against erosion and scour in accordance with Section 2.4; and
3. Have at least one door satisfying building code requirements for an exit door or primary means of escape, above the applicable elevation specified in Table 6-1, and capable of providing human ingress and egress during the design flood.

6.2.3 Limits on Human Intervention

Dry floodproofing measures that require human intervention to activate or implement prior to or during a flood shall be permitted only when all of the following conditions are satisfied:

1. The flood warning time (alerting potential flood victims of pending flood situation) shall be a minimum of 12 hours, unless the community operates a flood warning system and implements an emergency plan to ensure safe evacuation of flood hazard areas, in which case human intervention is

allowed only if the community can provide a minimum flood warning time equal to or longer than the cumulative
   (a) time to notify person(s) responsible for installation of floodproofing measures, plus
   (b) time for responsible persons to travel to structure to be floodproofed, plus
   (c) time to install, activate, or implement floodproofing measures, plus
   (d) time to evacuate all occupants from the flood hazard area;
2. All removable shields or covers for openings such as windows, doors, and other openings in walls shall be designed to resist flood loads specified in Section 1.6; and
3. Where removable shields are to be used, a flood emergency plan shall be approved by the authority having jurisdiction and shall specify, at a minimum, the following information: storage location(s) of the shields; the method of installation; conditions activating installation; maintenance of shields and attachment devices; periodic practice of installing shields; testing sump pumps and other drainage measures; and inspecting necessary material and equipment to activate or implement floodproofing. The flood emergency plan shall be permanently posted in at least two conspicuous locations within the structure.

6.3 WET FLOODPROOFING

Wet floodproofing shall be accomplished through the use of flood-damage-resistant materials and techniques that minimize damage to a structure during periods where the lower portion of the structure is inundated by floodwater. All materials in contact with floodwaters shall conform with the requirements of Section 5.

6.3.1 Wet Floodproofing Limitations on Use

Wet floodproofing of enclosed areas below the elevations listed in Table 6-1 shall be limited to

1. Category I structures;
2. Enclosures used solely for parking, building access, or storage;
3. Structures that are functionally dependent on close proximity to water; and
4. Agricultural structures not included in Category I structures that cannot be located elsewhere and that are used solely for agricultural purposes.

6.3.2 Wet Floodproofing Requirements

Wet floodproofing for flood events up to and including the design flood shall be accomplished by

1. Use of techniques that minimize damage to the structure associated with flood loads;
2. Meeting the requirements of Section 2.6 or Section 4.6, depending on the flood hazard area; and
3. Installation of utilities, including plumbing fixtures, in conformance with the requirements of Section 7.

7.0 UTILITIES

7.1 GENERAL

Utilities and attendant equipment shall not be located below the elevation specified in Table 7-1 unless

1. Specifically allowed in this section; and
2. Designed, constructed, and installed to prevent floodwaters, including any backflow through the system, from entering or accumulating within the components.

Utilities and attendant equipment shall be installed and anchored to resist flood forces.

Utilities and attendant equipment shall be permitted in areas of structures that are dry floodproofed. Utilities and attendant equipment shall not be mounted on, pass through, or be located along breakaway walls.

Elevated exterior platforms for utilities and attendant equipment shall be supported on piles or columns, or cantilevered from or knee braced to the structure. If piles or columns are utilized, they shall be adequately embedded to account for erosion and local scour around the supports.

7.2 ELECTRICAL SERVICE

7.2.1 Service Conduits and Cables

Electrical service conduits and cables below the Design Flood Elevation (DFE) shall be waterproofed or conform to the provisions of NFPA 70 *National Electrical Code,* Ref. [17], for wet locations. Underground service conduits and cables shall be buried to a depth sufficient to prevent movement, separation, or loss due to erosion and scour under design flood conditions.

7.2.2 Exposed Conduits and Cables

Electrical conduits and cables emerging from underground shall be designed, constructed, and installed to withstand flood-related loads, including the effects of buoyancy, hydrodynamic forces, and debris impacts. Waterproofing or protective enclosures shall be provided for nonwaterproof conduits and cables extending vertically to elevated structures. The enclosures shall be securely fastened to the structure; however, protective enclosures and electrical conduits and cables shall not be fastened to walls, enclosures, or structures intended to break away under flood conditions.

Electrical conduits and cables and protective enclosures installed below the elevations specified in Table 7-1 shall be sealed to prevent the entrance of floodwaters into electrical conduits and electrical service components.

7.2.3 Electric Meters

Electric meters shall be located above the elevation specified in Table 7-1 unless the connection between the meter and electric lines extending vertically from the meter is within a waterproof enclosure.

7.2.4 Disconnect Switches and Circuit Breakers

The main disconnect switch and all circuit breakers shall be located above and be accessible from above the elevation specified in Table 7-1. Switches and circuit breakers shall be located no more than 5 ft above the floor, or a platform shall be installed to provide access.

7.2.5 Electric Elements Installed Below Minimum Elevations

Where electrical conduits and cables are located below the elevation specified in Table 7-1, they shall be installed so as to drain water away from panelboards, controllers, switches, or other electrical equipment in accordance with NFPA 70 *National Electrical Code,* Ref. [17].

Where required to meet life safety provisions of the code, a minimum number of lighting circuits, switches, receptacles, and lighting fixtures operating at a maximum voltage of 120 volts to ground shall be permitted below the elevation specified in Table 7-1. Electrical wiring shall be suitable for submergence in water and only submersible-type splices shall be used. Switches, receptacles, and fixtures shall conform to the provisions of NFPA 70 *National Electrical Code,* Ref. [17], for wet locations and shall contain no fibrous components.

All circuits, switches, receptacles, fixtures, and other electrical components and equipment installed below the elevation specified in Table 7-1 shall be energized from a common distribution panel located above and accessible from above the elevation specified in Table 7-1.

In High Risk Flood Hazard Areas, main supply lines, meters, and other exterior electrical components installed below the elevation specified in Table 7-1 shall be installed on a non-breakaway vertical structural element on the landward, down slope, or downstream side of the structure.

TABLE 7-1. Minimum Elevation of Utilities and Attendant Equipment Relative to Base Flood Elevation (BFE) or Design Flood Elevation (DFE)

	Locate Utilities and Attendant Equipment Above[b]		
		Coastal High Hazard Areas and Coastal A Zones	
Structure Category[a]	Flood Hazard Areas	Orientation Parallel[c]	Orientation Perpendicular[c]
I	DFE	DFE	DFE
II	BFE + 1 ft or DFE, whichever is higher	BFE + 1 ft or DFE, whichever is higher	BFE + 2 ft or DFE, whichever is higher
III	BFE + 1 ft or DFE, whichever is higher	BFE + 2 ft or DFE, whichever is higher	BFE + 3 ft or DFE, whichever is higher
IV	BFE + 2 ft or DFE, whichever is higher	BFE + 2 ft or DFE, whichever is higher	BFE + 3 ft or DFE, whichever is higher

[a]See Table 1-1 for structure category descriptions.
[b]Locate utilities and attendant equipment above elevations shown unless otherwise provided in the text.
[c]Orientation of lowest horizontal structural member relative to the general direction of wave approach: parallel shall mean less than or equal to +20 degrees from the direction of approach; perpendicular shall mean greater than +20 degrees from the direction of approach.

7.3 PLUMBING SYSTEMS

For the purposes of this standard, plumbing systems shall include sanitary collection systems, rain runoff collection systems, sanitary facilities and plumbing fixtures, water supply systems (including hot water heaters and water conditioning equipment), and sewage disposal systems.

7.3.1 Buried Plumbing Systems

Where installed underground, piping and plumbing systems providing service to a structure shall be buried to a depth sufficient to prevent movement, separation, or loss due to flooding and erosion under design flood conditions.

7.3.2 Exposed Plumbing Systems

Plumbing systems and components emerging from underground shall be designed, constructed, anchored, and protected to withstand flood-related loads, including the effects of buoyancy, hydrodynamic forces, and debris impacts.

7.3.3 Plumbing Systems Installed Below Minimum Elevations

Plumbing systems and components, including plumbing fixtures, shall be elevated above the elevation specified in Table 7-1. Where plumbing systems and components have openings below the elevation specified in Table 7-1, the openings shall be protected with automatic backwater valves or other automatic backflow devices. Devices shall be installed in each line that extends below the DFE to prevent release of sewage into floodwaters and to prevent infiltration by floodwaters into the plumbing. Redundant devices requiring human intervention shall be permitted.

7.3.4 Sanitary Systems

Sanitary systems shall be designed to minimize infiltration of flood waters into the systems and discharges from the systems into floodwaters. Vents and openings shall be above the elevation specified in Table 7-1. Sanitary system storage tanks shall be designed, constructed, installed, and anchored to resist at least 1.5 times the potential buoyant and other flood forces acting on an empty tank during design flood conditions. Tanks and piping shall be installed to resist scour and erosion.

Sanitary systems that must remain operational during or immediately after the design flood or lesser floods shall be equipped with a sealed storage tank that is sized to store at least 150% of the anticipated sewage flow associated with occupancy during flood conditions and during subsequent periods of saturated soil when sewage will not percolate.

7.4 MECHANICAL, HEATING, VENTILATION, AND AIR CONDITIONING SYSTEMS

Fuel supply lines extending below the elevation specified in Table 7-1 shall be equipped with a float operated automatic control valve to shut off fuel supply when floodwaters rise above the elevation of the supply line.

In Coastal High Hazard Areas and Coastal A Zones, and when located on the exterior of a structure, mechanical or HVAC equipment and systems shall be located on the landward side of the structure, above the elevation specified in Table 7-1.

Ductwork and ductwork insulation shall be at or above the elevation specified in Table 7-1 unless designed, constructed, and installed to resist all flood-related loads and to prevent floodwater from entering or accumulating within the ductwork. Ductwork shall be designed and installed to prevent the entry of floodwaters into the structure.

7.4.1 Tanks

For tanks that are associated with utility service to a building or that are attached to or located under a building, tanks and tank inlets, fill openings, outlets, and vents shall be elevated above the elevation specified in Table 7-1, or they shall be designed, constructed, installed, and anchored to resist all flood-related and other loads during the design flood, or lesser floods, without release of contents into floodwaters or infiltration by floodwaters into the contents.

Tanks shall be designed, constructed, installed, and anchored to resist at least 1.5 times the potential buoyant and other flood forces acting on an empty tank during design flood conditions.

7.5 ELEVATORS

Unless otherwise permitted in this section, all elevator components shall be located above the elevation specified in Table 7-1. Elevator components located below the elevation specified in Table 7-1 shall be constructed of flood-damage-resistant materials and shall be capable of resisting physical damage due to design flood conditions.

Hydraulic elevators shall be permitted below the elevation specified in Table 7-1, but the electrical control panel, hydraulic pump, and tank shall be elevated above that elevation. Drainage shall be provided for the elevator pit. The hydraulic lines connecting to the hydraulic jack assembly shall be located to protect the lines from physical damage due to design flood conditions.

For traction elevator systems, the machine room containing the electric hoist motors and electrical control panel shall be located above the elevation specified in Table 7-1. Elevator components located in the hoistway below the elevation specified in Table 7-1 shall be protected from physical damage due to design flood conditions.

Where there is the potential for an elevator cab to descend below the elevation specified in Table 7-1 during a flood event, the elevator shall be equipped with controls that will prevent the cab from descending into floodwaters.

8.0 BUILDING ACCESS

8.1 GENERAL

Stairways and ramps that are located below the elevations specified in Tables 2-1 and 4-1 shall be designed and constructed to

1. Resist flood-related loads specified in Section 1.6 and to minimize transfer of flood-related loads to the structure and structure foundation; or
2. Break away during design flood conditions without causing damage to the structure, including the foundation.

Materials used for those portions of stairways and ramps that are located below the elevations specified in Tables 2-1 and 4-1, including items such as gates and doors, shall conform to Section 5.

Enclosures for stairways and ramps that extend below the elevations specified in Tables 2-1 and 4-1 shall conform to the requirements for enclosures in Sections 2.6 and 4.6.

Elevators shall conform to the requirements of Section 7.5.

9.0 MISCELLANEOUS CONSTRUCTION

9.1 GENERAL

Miscellaneous structures and construction governed by this section shall be designed to withstand all flood-related loads as defined in Section 1.6 of this standard. Structures shall be elevated as required in Section 2.3 or Section 4.4, or shall be designed and constructed in conformance with this section. Materials below the design flood elevation shall conform with Section 5. Utilities shall be governed by Section 7.

9.2 DECKS, PORCHES, AND PATIOS

Decks, porches, and patios may be attached to, or detached from, an adjacent structure, and are subject to the requirements in Section 9.2.1 or Section 9.2.2, as appropriate.

9.2.1 Attached Decks, Porches, and Patios

In flood hazard areas other than Coastal High Hazard Areas, Coastal A Zones, and other High Risk Flood Hazard Areas, decks, porches, and patios that are structurally connected to a structure shall be considered to function as a continuation of the structure and are allowed below the elevations specified in Table 2-1 provided the decks, porches, and patios conform to the foundation requirements of Section 1.5.3. If enclosed, the walls shall conform to the requirements for openings in enclosures in Section 2.6.

In Coastal High Hazard Areas, Coastal A Zones, and other High Risk Flood Hazard Areas, decks, porches, and patios that are structurally connected shall be considered to function as a continuation of the structure and shall be designed with the lowest horizontal member above the elevations specified in Table 4-1 and shall be elevated on piles, posts, columns, or walls serving as shear walls in conformance with Section 4, or cantilevered from or knee braced to the structure or the foundation system supporting the structure.

9.2.2 Detached Decks, Porches, and Patios

In flood hazard areas other than Coastal High Hazard Areas, Coastal A Zones, and other High Risk Flood Hazard Areas, decks, porches, and patios that are not structurally connected to a structure are permitted below the elevations specified in Table 2-1. Decks, porches, and patios shall conform to the foundation requirements of Section 1.5.3. If enclosed, the walls shall conform to the requirements for openings in enclosures in Section 2.6.

In Coastal High Hazard Areas, Coastal A Zones, and other High Risk Flood Hazard Areas, decks, porches, and patios that are not structurally connected to a structure are permitted below the elevations specified in Table 4-1 if designed in accordance with the requirements of Section 4.8. If enclosed, the walls shall conform to the breakaway wall requirements in Section 4.6.

9.3 GARAGES

9.3.1 Attached Garages and Carports

In flood hazard areas other than Coastal High Hazard Areas, Coastal A Zones, and other High Risk Flood Hazard Areas, garages and carports that are structurally connected to a structure shall be considered to function as a continuation of the structure. Garages and carports are permitted below the elevations specified in Table 2-1 provided the lowest floor of the garages and carports is at or above grade on at least two sides and the garage and carport walls meet the openings requirements of Section 2.6.

In flood hazard areas other than Coastal High Hazard Areas, Coastal A Zones, and other High Risk Flood Hazard Areas, garages and carports that are structurally connected to nonresidential buildings shall not have the lowest floor of the garages below grade on all sides unless they are floodproofed in conformance with Section 6.

In Coastal High Hazard Areas, Coastal A Zones, and other High Risk Flood Hazard Areas, garages that are structurally connected to a structure shall be considered to function as a continuation of the structure. Garages and carports below the elevation specified in Table 3-1 or Table 4-1 shall be constructed with breakaway walls in conformance with the requirements of Section 4.6. Garages and carports that are elevated shall be in conformance with Section 4.

In Coastal High Hazard Areas, Coastal A Zones, and other High Risk Flood Hazard Areas, garages and carports that are structurally connected to a structure shall not be permitted to have the lowest floor below grade on all sides.

9.3.2 Detached Garages and Carports

In flood hazard areas other than Coastal High Hazard Areas, Coastal A Zones, and other High Risk Flood Hazard Areas, garages and carports that are not structurally connected to a structure are permitted below the elevations specified in Table 2-1. Garages and carports shall conform to the foundation requirements of Section 1.5.3. If enclosed, the garage or carport walls shall conform to the requirements for openings in enclosures in Section 2.6.

In Coastal High Hazard Areas, Coastal A Zones, and other High Risk Flood Hazard Areas, garages and carports that are not structurally connected to a structure are permitted below the elevations specified in Table 4-1 if designed in accordance with the requirements of Section 4. If enclosed, the garage or carport walls shall conform to the breakaway wall requirements in Section 4.6.

9.4 CHIMNEYS AND FIREPLACES

In flood hazard areas other than Coastal High Hazard Areas, Coastal A Zones, and other High Risk Flood Hazard Areas, chimneys are permitted to extend below the elevation specified in Table 7-1. Chimneys extending below the elevation specified in Table 7-1 shall be vertically supported, shall be independent of the structure, and shall be designed and constructed to withstand all flood-related loads.

In Coastal High Hazard Areas, Coastal A Zones, and other High Risk Flood Hazard Areas, the base of the chimney or fireplace shall not extend below the elevation specified in Table 7-1. Where vertical support is required, chimneys shall be vertically supported on piles or column foundations and foundation embedment shall be at least as deep as the rest of the structure foundation or deeper where needed to support the chimney against flood-related and other loads. The chimney or fireplace system shall be designed to minimize transfer of flood-related and other loads or load combinations to the structure and structure foundation.

9.5 POOLS

In-ground and above-ground pools shall be designed to withstand all flood-related loads and load combinations.

Pools that are structurally connected to structures shall be designed to function as a continuation of the building or structure.

In Coastal High Hazard Areas, Coastal A Zones, and other High Risk Flood Hazard Areas, pools shall

1. Be elevated so that the lowest horizontal structural member is elevated as required in Section 4.4; or
2. Be designed and constructed to break away during design flood conditions without producing debris capable of causing significant damage to any structure; or
3. Be sited to remain in the ground during design flood conditions without obstructing flow that results in damage to any structure.

9.6 STORAGE TANKS

For tanks not associated with utility service to a building and that are not attached to or located under a building, above ground and subsurface storage tanks and tank inlets, fill openings, outlets, and vents that are located below the elevation specified in Table 7-1 shall be designed, constructed, installed, and anchored to resist all flood-related and other loads, including the effects of buoyancy, during flooding up to and including the design flood and without release of contents into floodwaters or infiltration by floodwaters into the tanks.

Tanks shall be designed, constructed, installed, and anchored to resist at least 1.5 times the potential buoyant and other flood forces acting on an empty tank during design flood conditions.

10.0 REFERENCES

All documents referenced in this standard are listed below with full titles and dates.

[1] American Society of Civil Engineers. 2002. Minimum Design Loads for Buildings and Other Structures, ASCE 7-02.

[2] American Concrete Institute. 2002. Building Code Requirements for Masonry Structures, ACI 530-02/ASCE 5-02/TMS 402-02.

[3] American Concrete Institute. 2002. Specification for Masonry Structures, ACI 530.1 02/ ASCE 6-02/TMS 602-02.

[4] American Concrete Institute. 2002. Building Code Requirements for Structural Concrete, ACI 318.

[5] American Society for Testing and Materials. 2000. Standard Test Method for Laboratory Compaction Characteristics of Soil Using Standard Effort (12,400 ft · lbf/ft^3 (600 kN · m/m^3)), ASTM D 698.

[6] American Society for Testing and Materials. 2000. Standard Test Method for Laboratory Compaction Characteristics of Soil Using Modified Effort (56,000 ft · lbf/ft^3 (2,700 kN · m/m^3)), ASTM D 1557.

[7] American Society for Testing and Materials. 1999. Standard Specification for Round Timber Piles, ASTM D 25.

[8] American Society for Testing and Materials. 1998. Standard Specification for Welded and Seamless Steel Pipe Piles, ASTM A 252.

[9] American Society for Testing and Materials. 2000. Standard Specification for Zinc (Hot-Dip Galvanized) Coatings on Iron and Steel Products, ASTM A 123/A 123M.

[10] American Society for Testing and Materials. 2000. Standard Specification for Zinc Coating (Hot-Dip) on Iron and Steel Hardware, ASTM A 153/A 153M.

[11] American Society for Testing and Materials. 2000. Standard Specification for Steel Sheet, Zinc-Coated (Galvanized) or Zinc-Iron Alloy-Coated (Galvannealed) by the Hot-Dip Process, ASTM A 653/A 653M.

[12] American Society for Testing and Materials. 1999. Standard Specification for General Requirements for Steel Sheet, Metallic-Coated by the Hot-Dip Process, ASTM A 924/A 924M.

[13] American Society for Testing and Materials. 2000. Standard Specification for Carbon Structural Steel, ASTM A 36/A 36M.

[14] American Society for Testing and Materials. 2000. Standard Specification for High-Strength Low-Alloy Columbium-Vanadium Structural Steel, ASTM A 572/A 572M.

[15] American Society for Testing and Materials. 2000. Standard Specification for High-Strength Low-Alloy Steel H-Piles and Sheet Piling for Use in Marine Environments, ASTM A 690/A 690M.

[16] American Forest & Paper Association. 2001. National Design Specification for Wood Construction.

[17] National Fire Protection Association. 2002. National Electrical Code, NFPA 70.

COMMENTARY

(This commentary is not part of ASCE/SEI 24-05. It is included for information purposes.)

C1.0 GENERAL

C1.1 SCOPE

The requirements of this standard are developed through a rigorous consensus process of the American Society of Civil Engineers and are intended to protect public safety and property. It is, in part, based on examination of damage after floods and is intended to meet or exceed the requirements of the National Flood Insurance Program (NFIP). Any conflicts or differences between this standard and other applicable regulations should be resolved such that compliance with NFIP requirements is equaled or exceeded. Certain terminologies of NFIP requirements not incorporated in the standard are provided in this commentary.

This standard applies to the design and construction of new structures located in flood hazard areas and subsequent work to such structures. Subsequent work includes additions that change the footprint of a structure and additions that change the number of floors. Subsequent work also includes alteration or replacement of foundations, an activity that may be undertaken specifically to improve a structure's resistance to flood loads and damage [e.g., to raise a substantially improved or substantially damaged structure above the design flood elevation (DFE)]. This standard also applies to work that is classified as substantial repair of damage of any cause and to improvements that constitute substantial improvement. In addition, the standard applies to foundations intended for placement or replacement of manufactured housing units.

This standard is not intended to apply to routine or minor repairs or improvements to a structure. However, it should be noted that, in some instances, repairs to or improvements of an existing structure may be sufficient for the authority having jurisdiction to classify the work as a "substantial improvement." This standard is intended to apply to restoration and repair of damage to a structure for which the authority having jurisdiction has made a determination of "substantial damage." In both instances, the authority having jurisdiction may require work to be in conformance with requirements for new construction. The authority having jurisdiction should be consulted to determine whether repairs or improvements in a flood hazard area require the use of this standard or other requirements commonly applied to new construction.

The provisions of this standard may not apply to certain historic structures (see Fig. 1-1). The authorities having jurisdiction over historic structures, such as the State Historic Preservation Office and local historic preservation authorities, should be consulted to determine if the proposed work allows the structure to retain its historic designation. If the structure will not retain its historic designation, the provisions of this standard apply. If the standard does not apply, owners of historic structures are encouraged to use the standard to the extent practicable in order to reduce the exposure of historically significant buildings, and their contents, to future flood damage.

Fig. 1-2 illustrates how the standard is organized. Designers should determine which sections are specifically applicable, based on the type of flood hazard area. Section 1 contains provisions that apply to all new construction and substantial improvements in all flood hazard areas. Additional provisions that are specific to the type of flood hazard area are set forth in Sections 2, 3, and 4. Section 2 applies to structures in flood hazard areas including High Risk Flood Hazard Areas other than Coastal High Hazard Areas and Coastal A Zones. In Section 3, High Risk Flood Hazard Areas are described and basic provisions are outlined, but design specifics for these areas are not specified. Section 4 applies to structures in Coastal High Hazard Areas and Coastal A Zones. The remaining sections, Sections 5, 6, 7, 8, and 9, apply to all new construction and to substantial improvements in all flood hazard areas. This standard is not intended to preclude construction of piers, docks, wharves, and other water-dependent (functionally dependent) structures.

C1.2 DEFINITIONS

A Zone—area within the Special Flood Hazard Area that is not subject to high velocity wave action.

Breakaway Wall—by using the phrase "under base flood or lesser flood conditions" in the definition, this standard is not requiring the design and construction of breakaway walls that will fail under any applied load that is less than that occurring during the base flood. The load requirements of ASCE 7 (Standard Ref. [1]) must be met. ASCE 7 requires breakaway walls to withstand the design lateral wind loads and seismic loads, but to fail under base flood conditions. In no instance does ASCE 7 permit breakaway walls that will fail at lateral loads less than 10 psf.

Coastal A Zone—Coastal A Zone designation is used to facilitate application of load combinations contained in ASCE 7 (Standard Ref. [1]). Coastal A Zones are areas where breaking wave heights greater than or equal to 1.5 ft during the base flood are expected; they lie landward of zones designated by FEMA as V Zones (Fig. C1-1) and landward of an open coastal shoreline where V Zones have not been mapped (Fig. C1-2). Postflood field investigations and laboratory testing indicate significant structural damage occurs in these areas and is associated with wave impacts. Some Coastal A Zones experience erosion that is similar to that experienced in V Zones, and this standard requires design and construction similar to that required in V Zones.

The Coastal A Zone is not delineated on flood hazard maps prepared by FEMA, but is a zone where wave forces and erosion potential should be taken into consideration. In order for a Coastal A Zone to be present, two conditions are required: (1) a stillwater depth greater than or equal to 2 ft; and (2) breaking wave heights greater than or equal to 1.5 ft. Note that the stillwater depth requirement is necessary, but is not sufficient by itself, to render an area a Coastal A Zone. Many A Zones will have stillwater flood depths in excess of 2 ft, but will not experience breaking wave heights greater than or equal to 1.5 ft and therefore should not be classified as Coastal A Zones.

Design Flood Elevation (DFE)—based on the Design Flood, the DFE is the higher of the Base Flood Elevation (BFE) shown on FIRMs prepared by FEMA or the flood elevations shown on the map adopted by a community. Communities may elect to adopt flood elevations that are higher than those determined by FEMA. Such higher flood elevations may be adopted for a number of reasons, for example, to show future conditions (assuming predicted upland development), to reflect historic events that exceeded the 1%-annual chance flood, or to incorporate freeboard as an additional factor of safety to reflect local conditions (such as high debris loads not reflected in BFEs).

Development—any man-made change to improved or unimproved real estate, including but not limited to structures, permanent storage of materials, mining, dredging, filling, grading, paving, excavations, drilling operations, and other land disturbing activities.

FEMA—federal agency, or part of a federal agency, that, in addition to carrying out other activities, oversees the administration of the National Flood Insurance Program. When the U.S. Department of Homeland Security was organized in 2003 the Federal Emergency Management Agency was included as the Emergency Preparedness and Response Directorate. This Directorate, still referred to as FEMA, is charged with preparing for, mitigating the effects of, responding to, and recovering from major domestic disasters, both natural and man-made.

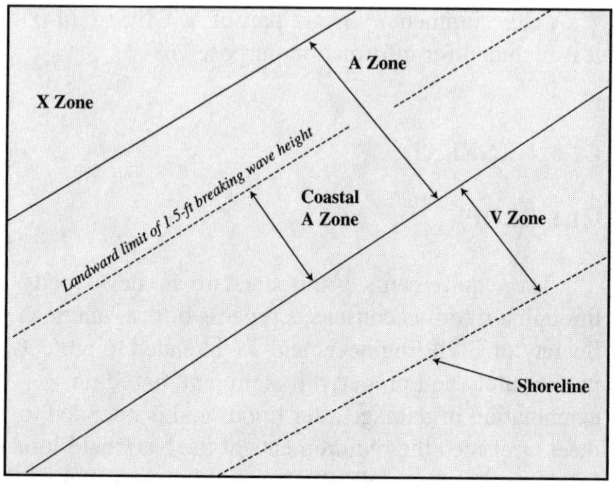

FIGURE C1-1. Coastal A Zone Where a Coastal High Hazard Area (V Zone) is Present.

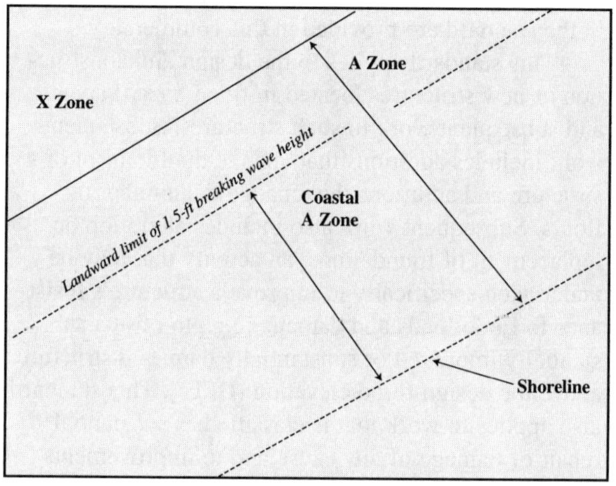

FIGURE C1-2. Coastal A Zone Where a Coastal High Hazard Area (V Zone) is Not Present.

Flood Hazard Boundary Map (FHBM)—first flood risk map prepared by FEMA for a community, which identifies flood hazard areas based on the approximation of land areas in the community having a 1% or greater chance of flooding in any given year.

Floodproofing—floodproofing measures outlined in this standard may apply to new construction or retrofit of existing structures. Under the NFIP, floodproofing that is intended to protect areas within a structure that are below the Base Flood Elevation is allowed only for nonresidential buildings.

Flood Insurance Rate Map (FIRM)—official map of a community, on which FEMA has delineated both the special hazard areas and the risk premium zones applicable to a community. Many communities have multiple panels comprising the FIRM, and panels may have different effective dates. Users should verify that the most current effective map is used in determining flood hazard areas, floodway areas, and base flood elevations. FEMA's online Map Store is accessible at http://www.fema.gov.

Flood Insurance Study (FIS)—examination, evaluation, and determination of flood hazards and, if appropriate, corresponding water surface elevations, or an examination, evaluation, and determination of mudslide (mudflow) and/or flood-related erosion hazards, and a report prepared by FEMA of the flood hazards studied for a community. The FIS typically includes a section on discharges of streams studied, water surface elevation profiles for floods of different frequencies, and floodway data tables.

Floodplain Management Ordinance—ordinance adopted by the authority having jurisdiction to regulate development in flood hazard areas. The ordinance addresses all development, including development activities that are outside the scope of this standard.

Freeboard—additional height used as a factor of safety in setting the minimum elevation of a structure, or floodproofing measures applied to a structure, to compensate for factors that may increase flood heights and for uncertainties inherent in determining flood frequencies and flood elevations.

Lowest Adjacent Grade—lowest elevation of the natural or regraded ground surface, or structural fill, at the location of a structure.

Manufactured Housing—structure that is transportable in one or more sections, built on a permanent chassis and constructed to Federal Mobile Home Construction and Safety Standards and rules and regulations promulgated by the U.S. Department of Housing and Urban Development. The term does not include a "recreational vehicle."

Mudslide—definition of mudslide is consistent with the NFIP use. However, it is intended to include only classes of flow that are technically recognized as mudflow and mud flood. Mudslides (mudflows and mud floods) occur where sediment concentrations are between 0.20 and 0.55 (by volume) or between 0.41 and 0.79 (by weight). Where sediment concentrations are less than those limits, the condition is a flood. Where sediment exceeds those limits, the condition is a landslide. Landslides are not intended to be covered in this standard since they are not directly flood related.

National Flood Insurance Program (NFIP)—federal program to provide relief from the impacts of flood damages in the form of federally backed flood insurance that is available to participating communities, contingent upon nonstructural flood loss reduction measures embodied in local floodplain management regulations adopted by a community. The program is administered by FEMA.

Prolonged Contact with Floodwaters—definition used in this standard for coastal flooding differs from the definition used in NFIP Technical Bulletin 2-93 *Flood-Resistant Materials Requirements for Buildings Located in Special Flood Hazard Areas,* Ref. [C1], which refers to inundation for longer than 72 hours. Coastal flooding usually is of short duration; therefore materials are unlikely to be inundated for durations longer than 12 hours. Where durations are known or determined through analyses to be other than the durations of inundation specified in the definition, it is appropriate to consider those known or determined durations when specifying materials for use below the design flood elevation. For example, in areas where coastal flooding has previously caused ponding, specification of materials resistant to inundation durations longer than 12 hours may be appropriate.

C1.3 IDENTIFICATION OF FLOOD HAZARD AREAS

The determination as to whether or not a structure lies within the Special Flood Hazard Area begins by determining the map that has been adopted by the community. While most communities adopt flood hazard maps prepared by FEMA, other maps that show larger affected areas and higher Design Flood Elevations may be used for regulatory purposes. Flood Insurance Rate Maps (FIRM) or Flood Hazard Boundary Maps (FHBM) are produced or approved by FEMA as part of the NFIP's nationwide floodplain identification efforts. See Ref. [C2] for more information on using FIRMs to identify flood hazard areas. Ref. [C3] provides a comprehensive overview of floodplain mapping and floodplain management issues.

Flood hazard areas on the FEMA maps are referred to as "special flood hazard areas." Designations for most of these areas of special flood hazard begin with the letter A or V followed by another letter or number, indicating that flood elevations and flood hazard factors have been determined. If A or V is not followed by another letter or number, flood elevations and other flood hazard factors have not been determined. When a structure will be located in an unnumbered zone, consult with the authority having jurisdiction for additional guidance.

In flood hazard areas for which detailed elevation information is not shown on a community's FIRMs, information may be obtained from technical studies produced in conjunction with FEMA mapping or from other sources, such as the U.S. Army Corps of Engineers, the U.S. Natural Resources Conservation Service (formerly known as the U.S. Soil Conservation Service), the U.S. Geological Survey, and state or local floodplain management agencies and transportation agencies.

In the absence of adequate information from those organizations listed above, practitioners in hydrology, hydraulics, or coastal engineering can be consulted to determine the flood hazard potential for a site, with an analysis of site characteristics, with engineering calculations performed with a variety of FEMA-approved models (listed by FEMA at http://www.fema.gov), and with historical flood records (if available). Practitioners in hydrology and hydraulics can calculate riverine flood elevations and floodways. Practitioners in coastal engineering can calculate storm surge elevations and storm-induced erosion with a variety of approved models. It should be noted that in areas where infrequent but extreme flood events occur, the lack of sufficient historical data makes such determinations difficult and reduces the reliability of flood frequency analyses.

In flood hazard areas subject to flooding from riverine sources (noncoastal), all zones begin with the letter A. Many waterways that have detailed elevations and flood hazard factors have designated "floodways." Floodways generally are areas where the greatest conveyance occurs, thus they often have higher velocities. Development within floodways can increase flood elevations; thus additional engineering evaluation of the impact of proposed development is required (see Section C1.5.4).

In flood hazard areas subject to flooding from coastal sources, zones beginning with an A represent areas that have a 1%, or greater, probability of flooding during any year, but which are not expected to experience wave heights of 3 ft or more or wave runup depths of 3 ft or greater (high velocity wave action). Zones beginning with a V represent areas that have a 1%, or greater, probability of flooding during any year, and where high velocity wave action is anticipated during the 1% flood. Zones beginning with a V are also referred to as *Coastal High Hazard Areas;* however, high velocity wave action can occur in lake or riverine floodplains under certain circumstances as well.

Coastal A Zones, where wave forces and the potential for erosion are present, are described in Section C2.1.

The national floodplain management standard administered by the NFIP and reflected on maps prepared by FEMA is the 1%-annual chance flood (commonly called 100-year flood). This event has a 1% chance of being equaled or exceeded in any given year. The NFIP refers to this flood as the "base flood." A "base flood elevation" (BFE)—the water surface elevation associated with the base flood (relative to the National Geodetic Vertical Datum [NGVD], the North American Vertical Datum [NAVD], or another datum)—is identified on many of the NFIP's maps.

Designers are reminded to exercise caution regarding the use of the term "100-year flood." The public frequently misinterprets the term, believing that the occurrence of a 100-year flood precludes another flood of similar magnitude for 100 years. Wherever possible, the term "1%-annual chance flood" should be used instead of "100-year flood."

All communities that participate in the NFIP (approximately 20,000 communities as of 2003) are required to adopt floodplain management regulations that meet or exceed the NFIP requirements. Designers should be aware, however, that many states and communities enforce more restrictive requirements. More restrictive requirements may be found in virtually all aspects of local regulations. Some of the more common are the following: adopting a higher flood elevation than the NFIP's base flood; requiring construction to be elevated higher than the NFIP's requirements; requiring engineered foundations; or imposing other more restrictive siting, development, or construction standards. The authority having jurisdiction should be consulted to determine the applicable requirements for construction in floodplain areas.

For the purposes of this standard, the terms "design flood" and "design flood elevation" will be used to refer to the locally adopted regulatory flood and its associated water surface elevation. If the authority having jurisdiction regulates to NFIP minimum requirements, then the design flood and design flood elevation will be identical to the base flood and base flood elevation. If the authority having jurisdiction regulates to a higher flood elevation, then the design flood and design flood elevation will be greater than the base flood and base flood elevation.

It should be noted that—in coastal areas—the design flood elevation and the base flood elevation will include the effects of waves; i.e., the DFE and BFE will be established at the wave crest elevation or wave runup elevation, not the stillwater elevation. The designer is referred to Section 3 for more information on this topic.

Even if the authority having jurisdiction has adopted NFIP minimum requirements, it may still be a prudent practice to design and construct a structure to a higher elevation and more conservative standard than that required. In some cases, V zones have not been mapped where high velocity wave action is known to occur (e.g., along the shorelines of the Great Lakes). In some cases mapping efforts are hindered by the fact that it is not possible to assign an accurate exceedance frequency to historical flood events, particularly in cases where an area is subject to unique hazards (alluvial fan flooding, mudflows, ice jams, debris blockage of culverts and bridges, etc.). As a consequence, many hazard areas have not yet been mapped—but the hazards are known to exist. Design and construction in such instances should rely on historical data and take all known or likely hazards into consideration.

C1.4 IDENTIFICATION OF FLOODPRONE STRUCTURES

C1.4.1 General

Flood hazard determinations should be based on the most recent and accurate maps, studies, and data available. Therefore, the community should always be contacted to obtain the latest information. However, the designer is warned that even the latest maps and studies may be incomplete, inaccurate, or obsolete. In instances where maps, studies, or other information are known by the designer to be incomplete, inaccurate, or obsolete, the community may be able to assist the designer in obtaining and reviewing flood hazard data from other sources, or may be able to provide a listing of other sources that the designer can contact.

This standard applies to all structures that are located, in whole or in part, in flood hazard areas. A structure that "straddles" a line between flood hazard areas, or between a flood hazard area and an area not designated as subject to flooding, must meet the most restrictive requirements for either area. Designers are cautioned about imbuing a designation on a map with a higher degree of precision than intended. This caution is made with regard to several aspects of flood hazards and flood hazard maps, including but not limited to the following:

1. Determining flood hazard areas and flood elevations involves a great deal of analysis of historical records and/or modeling; the modeler has numerous assumptions to make and parameters to determine based on judgment.
2. The 1%-annual chance flood is a statement of probability that is, for the most part, a function of past flood events; as the record of flood events at any given location accumulates and especially as large floods occur, a statistical analysis of that record likely will change the flood flows and elevations.
3. In riverine areas, changes in land use over time alter the rainfall-runoff relationship and contribute to increases in flood elevations.
4. In coastal areas, modeled results are highly sensitive to the shoreline topography used, yet topography may change over time due to erosion or development.
5. The base maps on which many flood hazard areas are delineated are not of sufficient scale to capture all ground variations; an area may not be shown as subject to flooding, but the elevation of the ground may, in fact, be lower than the design flood elevation.
6. The scale of most flood hazard maps is such that the width of lines can be a factor in determining whether a structure is "in" or "out" of the flood hazard area; designers are cautioned to be liberal in their determinations in order to protect property and public safety.

C1.4.2 Consideration for Flood Protective Works

Thousands of miles of levees and other flood protective works are in existence. Many (but not all) have been designed, constructed, and maintained so as to provide protection against flood conditions, although perhaps not the design flood as defined by this standard. Many flood protective works were built for agricultural, emergency, or other purposes, and may provide protection against only smaller floods (e.g., 5- 15-year flood frequencies). Many will not provide protection during the base flood or the design flood, and have not been recognized and credited as providing flood protection during the mapping of flood hazard areas by FEMA. Unfortunately, levees and other similar works—even small and poorly maintained ones—often provide a false sense of security. When these works fail or are overtopped by floods that exceed their crest elevation, rapid inundation and high velocity flows often result. Even if the works do not fail or overtop during a flood, if not appropriately designed and maintained, they may cause flooding by trapping rainfall, runoff, and other waters behind them. See Chapter 4 of Ref. [C4] for more information on areas behind unsafe or inadequate levees.

Only certain works that have been designed, constructed, and maintained to rigorous standards have been recognized by FEMA as providing protection during the 1%-annual chance flood event and have been considered in floodplain mapping. When recognized by FEMA, the area protected by major flood control works is removed from the special flood hazard area, and flood insurance is not required.

Nevertheless, the area may be identified on the flood insurance rate map as having residual flood risk from floods greater than the 1%-annual chance flood.

However, even certified flood protective works may not render the area behind them as being safe for the design and construction of structures at grade. For example, the integrity of certain flood protective works may have been reduced by flooding, erosion, poor maintenance, or other events since the floodplain was mapped. More commonly, areas protected by even well-designed and constructed works are known to suffer from drainage problems, especially during extreme flood conditions.

Prudent design and construction of structures in proximity to flood protective works will entail the review of all available maps, studies, and other information pertaining to the works and its condition, and any known drainage or ponding problems in the area.

C1.4.3 Classification of Structures

Table 1-1 of the standard classifies structures according to importance and potential threat to human life. The table has been taken directly from ASCE-7 (Standard Ref. [1]), and it may be different from the classifications in building codes. For the purposes of this, only structures that represent a low hazard to human life and that pose little threat to other structures (during design flood conditions) should be classified as Category I.

Note that while agricultural facilities are classified as Category I structures and are subject to the requirements of this standard, certain agricultural facilities may be constructed with their lowest floors below the design flood elevation. For example, FEMA recognizes that wet floodproofing may be appropriate for some agricultural structures—farm storage structures, grain bins, corn cribs, and general purpose barns—even though these structures are not used for those purposes specifically identified in the NFIP regulations for enclosures below the BFE (parking, building access, storage) as being suitable for wet floodproofing. The NFIP will allow these types of agricultural structures to be wet floodproofed upon issuance of a variance from the community. NFIP Technical Bulletin 7-93, Ref. [C5], provides more information on this subject.

C1.5 BASIC DESIGN AND CONSTRUCTION REQUIREMENTS

C1.5.1 General

It is essential that the designer understand the nature and magnitude of loads that will act on structures during design flood conditions. Furthermore, it is essential that appropriate load combinations be applied, especially for coastal areas (Coastal High Hazard Areas and Coastal A Zones) and other areas around large bodies of water. Design wind loads may act simultaneously with design flood loads in such areas. This is usually not the case in riverine areas; however, in riverine floods of long duration, where flood water depths and wide areas of water or low geography create long fetches, wind-driven waves may develop. In these areas, wave action may be important and should be considered in design and construction.

Proper identification and/or calculation of DFEs and velocities are, of course, required before structure elevation, foundation type, and other aspects of design can be considered. Design flood depths and velocities must be specified with a reasonable degree of accuracy; otherwise, flood forces computed during the design process will be unreliable. If flood hazard maps and studies do not specify flood elevations (e.g., as in the case of unnumbered A zones) or flood discharge/velocity information, other sources of floodplain information should be consulted. Agencies listed in Section C1.3 should be contacted. In the absence of flood elevation and other data from FEMA or other agencies, registered professional engineers having the required expertise in hydrology and hydraulics should be consulted.

Siting of structures in flood hazard areas involves land use and regulatory issues. The designer should consult the local jurisdiction and other agencies with zoning, land use, and regulatory authority, to ascertain local and state siting restrictions or conditions. Many jurisdictions or governing authorities may prohibit construction or other development activities in certain portions of the floodplain; many will impose specific requirements on siting, design, and construction. For example, some jurisdictions may require that new construction not cause any of the following to occur: (1) alter the flood path; (2) increase the flood discharge: (3) increase the flood velocity; (4) increase the flood elevation; (5) increase the area of flood inundation; or (6) reduce available wetlands.

C1.5.2 Elevation Requirements

The minimum required elevation of structures (lowest floor, dry floodproofing measures, and location of utility equipment) is a function of two factors: the type of flood hazard area and the structure category. Tables that specify the minimum elevations are found in Section 2 (flood hazard areas other than Coastal High Hazard Areas and Coastal A), Section 4

(Coastal High Hazard Areas and Coastal A), Section 5 (flood damage resistant materials), Section 6 (floodproofing), and Section 7 (utilities and attendant equipment). In general, lowest floor elevations are higher in Coastal High Hazard Areas and Coastal A Zones, and also vary as a function of whether the lowest horizontal structural member is parallel to or perpendicular to the direction of wave approach.

Structure Category IV structures are essential facilities and warrant a higher level of protection than other structures. The minimum elevation tables specify that Category IV structures are to be elevated to the BFE plus 2 or 3 ft, or to the DFE, whichever is higher. Similarly, Category III structures, which represent a substantial hazard to human life in the event of failure, also warrant a higher level of protection; the elevation tables specify the minimum elevation to be BFE plus 1 or 2 ft, or the DFE, whichever is higher. Category II structures, which include one- and two-family homes and townhouses, are specified as having the lowest floor located at a minimum elevation of BFE plus 1 or 2 ft, or the DFE, whichever is higher. The minimum elevation requirement for Category I structures is the DFE (which is defined as the BFE or the DFE, whichever is higher).

C1.5.3 Foundation Requirements

C1.5.3.1 Geotechnical Considerations

Foundation design should be based on an accurate identification of underlying soil and rock properties at the site of a proposed structure. Although not required by many floodplain management ordinances or other applicable construction standards, geotechnical investigations should be conducted at any construction site in a flood hazard area, where available geotechnical data are insufficient for design purposes. Foundation design should take all potential impacts of soil saturation (especially during conditions of long duration flooding), consolidation, movement, expansion, and erosion into account, including the effects of long-term erosion that may occur at a site.

C1.5.3.2 Foundation Depth

In addition to the guidance contained in Sections 3 and 4 of this standard, the designer can refer to the body of literature dealing with flooding and flood-induced erosion, for example, Refs. [C6], [C7], [C8], and [C9]. The designer should consult professionals experienced in hydraulic and sediment transport analyses in the geographic area of interest.

C1.5.3.3 Foundation Walls

Use of foundation walls below the DFE will be permitted only where the walls are designed and constructed to resist flood-related forces acting on those walls so as not to jeopardize the stability or integrity of the structure. In most instances, use of load bearing walls below the DFE in flood hazard areas susceptible to high velocity flow, high velocity wave action, or other destructive flood forces will likely result in significant damage to, or total destruction of, a structure during design flood conditions. Load bearing foundation walls designed and constructed to withstand high velocity flows, high velocity wave action, or other destructive flood forces will not generally be affordable or cost-effective for construction less than three stories above the DFE.

Foundation walls that enclose areas below the DFE, also called solid perimeter walls or crawlspace foundations, must meet design specifications in Section 2.6. Solid perimeter wall foundations that are filled with engineered compacted fill topped with a slab (stem-wall foundations) are not subject to the specifications of Section 2.6.

In the case where foundation walls do not enclose an area below the DFE, it is recommended that the walls be oriented parallel to the direction of flood flow. This will reduce hydrodynamic loads, debris impact loads, and other flood-related loads that will act during design flood conditions.

C1.5.3.4 Piers, Posts, Columns, or Piles

Although this standard permits the use of piers, posts, and columns to elevate structures above the DFE, many failures of these foundation elements have been observed following coastal flood events, especially in areas where designs have not adequately accounted for potential erosion. The designer is cautioned that the use of piles may be more appropriate in coastal areas. In flood hazard areas other than Coastal High Hazard Areas and Coastal A Zones, use of piers, posts, columns, or piles, rather than placement of fill to achieve the appropriate minimum elevation, will minimize reduction of flood storage and diversion of floodwaters and drainage onto other properties.

C1.5.4 Use of Fill

In many areas of the country, fill is commonly used to meet the minimum elevation requirements in flood hazard areas, although structural fill is not allowed under structures in Coastal High Hazard Areas and Coastal A Zones. In most cases, fill is placed, compacted, and shaped to receive a slab-on-grade foundation. The placement of fill reduces flood storage and may increase local velocities; however, an engineering analysis of that effect is required under the NFIP only if the fill encroaches into the floodway. The designer should examine other site conditions when determining the feasibility of the use of fill and the design of the fill. In addition to specifying the type of fill materials and the appropriate compaction specifications, designers should consider specifying fill slopes that are no steeper than 2 horizontal to 1 vertical to minimize erosion, protect the foundation, and facilitate maintenance. Vegetative cover that is resistant to flowing water can be specified for fill slopes subject to low velocities. Slopes subject to high velocities should be protected by other means.

Designers should also be aware that the documentation of elevations required to demonstrate compliance (and for owners to obtain federal flood insurance) require the determination of the elevation of the Lowest Adjacent Grade. A structure that has the Lowest Adjacent Grade above the Base Flood Elevation may be determined by FEMA to be out of the mapped special flood hazard area.

C1.5.5 Anchorage and Connections

Designers should be aware that combinations of materials and connection types are possible and that possible failure modes using those connection types should be considered. The continuous load path concept is important enough to emphasize that the load path is made through the connections used in the building frame. A failure in any of the possible modes potentially weakens the entire structure, depending on how the connections are loaded (shear, tension, compression, torsion, etc.). It is also important to point out that loads occur in two directions, vertical and lateral (horizontal), and load combinations must be applied in both directions.

Bolts are appropriate for wood-to-wood, wood-to-steel, and wood-to-concrete connections. Steel-to-steel and steel-to-concrete connections may be made using bolted connections or welded connections. Connections using bolts or welds should be designed using the appropriate load combination strategy and may be designed for tensile, compression, and shear loads. Connections should be made such that the full capacity of the connection can be developed. Bolt holes should be located with sufficient clearances to minimize wood crushing, concrete cracking, steel tearing, and other failure modes. Welding on plates of sufficient thickness should be conducted with welding electrodes of sufficient capacity to develop the full connection capacity.

C1.6 LOADS IN FLOOD HAZARD AREAS

Extensive commentary on design loads and load combinations is contained in ASCE 7 (Standard Ref. [1]).

C1.6.1 General

While most loads acting on structures (such as wind loads) are typically on the order of tens of pounds per square foot, some flood-related loads (loads from high velocity waters, wave action, debris impacts, and ice) may exceed those typical loads by a factor of 10 to 100, or more. Extreme loads such as these make design and construction impractical and cost-prohibitive for structures that are not elevated above these hazards on a pile or column foundation that allows waves and flows to pass beneath the elevated structure.

C2.0 BASIC REQUIREMENTS FOR FLOOD HAZARD AREAS THAT ARE NOT IDENTIFIED AS COASTAL HIGH HAZARD AREAS AND COASTAL A ZONES

C2.1 SCOPE

This section defines basic requirements for design and construction of new construction and substantial improvement of existing structures located in flood hazard areas, including High Risk Flood Hazard Areas but not including Coastal High Hazard areas (V Zones) and Coastal A Zones. Therefore, it applies in flood hazard areas commonly called A Zones, numbered A Zones, AE Zones, and unnumbered A Zones, AO Zones, AH Zones, and AR Zones. Included are all floodplains along rivers, creeks, streams, and lakes, and in some A Zones that are inland of the Coastal A Zone. Even if a site is mapped as an A Zone on a flood hazard map, it is strongly recommended that the designer verify the lack of high velocity wave action, high velocity flows, erosion, debris effects, icejams, and other extreme flood hazards before applying the provisions of this section. See Section 3 for additional requirements for High Risk Flood Hazard Areas, and Section 4 for requirements related to Coastal High Hazard Areas and Coastal A Zone. See Sections 5 through 9 for requirements related to Flood-Resistant Materials, Dry and Wet Floodproofing, Utilities, Building Access, and Accessory Structures.

C2.2 DEVELOPMENT IN FLOODWAYS

The floodway is generally considered to be the most hazardous portion of a riverine floodplain area—

it conveys the greatest portion of the flood flow. Encroachment by construction, fill, or other development activities into the floodway can cause flood elevations to increase and can have other adverse consequences.

Regulatory floodways are often designated by the authority having jurisdiction as areas within which construction, fill, or development activities will not be permitted unless hydrologic and hydraulic analyses demonstrate the proposed activity will result in no decrease in effective conveyance and no increase in flood levels during the design flood. Computer programs such as the U.S. Army Corps of Engineers' HEC-2, Ref. [C10(b)], or HEC-RAS, Ref. [C10(c)], provide for such computations, so that carefully designated floodways will allow for sound utilization of flood hazard areas for development. The authority having jurisdiction should be contacted for a list of approved models and procedures.

Many of the flood insurance studies published by FEMA include regulatory floodways established by agreement between FEMA and the community. Since many Flood Insurance Rate Maps (FIRMs) have floodways designated, new computations are generally not required provided proposed development activities do not encroach into the designated floodway. However, lack of a designated regulatory floodway on a FIRM or Flood Hazard Boundary Map does not mean that a floodway does not exist—designers should check with the authority having jurisdiction and with FEMA before assuming flood hazards are uniform throughout the floodplain. In instances where regulatory floodways have not been designated, but where flood elevations have been determined, Section 60.3(c)(10) of the National Flood Insurance Program (NFIP) regulations requires communities to prohibit new construction, substantial improvements, or other development (including fill) in Zones A1-30 and AE shown on the community's FIRM, unless it is demonstrated that the cumulative effect of the proposed floodplain encroachments will not increase the water surface elevation of the base flood more than 1 ft at any point within the community. Some communities are even more restrictive—requiring demonstration that the cumulative effect of the proposed development will not increase the water surface elevation of the base flood at any point within the community. Some communities may require conveyance and/or storage compensation or other mitigation for any construction in the floodway.

Designers should check with the authority having jurisdiction to determine which development and construction activities are permitted in floodway areas, and to determine any limitations, evaluations, or special permit conditions associated with those activities. The designer should be advised that some states and communities severely restrict construction and development in floodways, and may prohibit residential construction entirely. For example, the State of Michigan prohibits all residential construction in floodway areas to ensure that the floodway remains uninhabited.

C2.3 ELEVATION REQUIREMENTS

A basic requirement for design and construction in flood hazard areas is that a structure should be capable of resisting flood forces and minimizing flood damages. This is usually accomplished by elevating the lowest floor (including basement) to or above the design flood elevation (DFE). NFIP regulations generally require elevation of the top of the lowest floor to or above the Base Flood Elevation (BFE). This standard generally requires elevation to or above the DFE. The design flood is the greater of the base flood or the flood corresponding to the flood hazard area shown on the community's flood hazard map. Therefore, this standard is sometimes more restrictive than NFIP requirements. Note, however, that there may be significant reduction in damage as well as savings in flood insurance premiums by elevating to higher than NFIP-required elevations. Note also, elevation to the BFE provides no safety factor against flooding. A safety factor can be incorporated by elevating above the BFE (i.e., by including freeboard).

The designer is cautioned that freeboard requirements included in Tables 2-1, 4-1, 5-1, 6-1, and 7-1 do not provide the same level of protection in all geographic areas. For example, 1 ft of freeboard may be equivalent to a 200-year flood in one area and a 500-year flood in another. Designers may wish to investigate the issue thoroughly before accepting this standard's minimum freeboard requirements—additional freeboard might be desirable.

In flood hazard areas other than Coastal High Hazard Areas (or other extreme flood loads associated with debris, ice, high velocity flows, etc.), elevation of nonresidential structures (and nonresidential portions of mixed-use structures) above the DFE is not always required by this standard. Flood damage reduction can sometimes be accomplished through floodproofing in such areas where the lowest floor is below the DFE. Check with the authority having jurisdiction for any specific floodproofing requirements and restrictions. The designer is also referred to Sections 5 and 6 for guidance and requirements related to flood-resistant materials and floodproofing. For floodproofed parking that is below grade (allowed only when associated with nonresidential structures), designers are referred to NFIP Technical Bulletin 6-93, *Below-Grade Parking Requirements for Buildings Located in Special Flood Hazard Areas,* Ref. [C11]. The designer is cautioned, however, that locating the lowest floor below the DFE may result in increased construction and maintenance costs, may lead to significant damage if the DFE is exceeded by a larger flood, and may result in substantially higher flood insurance premiums.

Design of structures situated near a flood hazard area but outside the flood hazard area should also make use of floodproofing and other flood-resistant practices. In particular, basement excavation and construction at a level below the DFE should be accomplished using appropriate design and construction.

It should be noted that the lowest floor of all new or substantially improved structures shall be at or above the elevations in Table 2-1, except for enclosed areas used solely for parking, building access, and storage, and except for those areas floodproofed in accordance with Section 6. These requirements prohibit basements in residential construction. However, NFIP regulations permit engineered basements beneath residential structures in very limited circumstances: these "basement exceptions" are granted based on technical criteria requiring the community to demonstrate that the Special Flood Hazard Area, as identified on the Flood Insurance Rate Map and in the Flood Insurance Study, is subject to shallow and low velocity flooding and that there is adequate flood warning time to ensure that all residents are notified of impending floods. As of 2003, only 54 communities have received basement exceptions from the NFIP. It is not the intent of this standard to prohibit residential basements in communities that have received basement exceptions from the NFIP. The floodproofing provisions in Section 6 of this standard are appropriate for consideration when designing a floodproofed residential basement, even though Section 6 primarily addresses nonresidential buildings.

C2.4 USE OF FILL

C2.4.1 Structural Fill

Structural fill should be used only in flood hazard areas not susceptible to high velocity wave action and other forces capable of eroding the fill. Structural fill should be suitable for its intended use, and should not lead to unacceptable levels of expansion, consolidation, or movement; structural fill should be granular and free-draining, wherever feasible. Structural fill used for foundation support and protection should be properly designed, constructed, and protected. Guidance on the protection of earth slopes to resist flood-related erosion from wave action and/or high flow velocities is available in several U.S. Army Corps of Engineers' publications, including Refs. [C6], [C12], and [C13]. For low flood velocities (5 ft/sec, or less) adjacent to structural fills, fill and slope protection is normally achieved by creating slopes of less than 1 vertical to 1.5 horizontal and providing appropriate vegetative cover; protection against moderate flood velocities (5–8 ft/sec) will require even flatter slopes and/or the use of stone or other materials. Use of structural fill for flow velocities greater than 8 ft/sec may not be feasible.

Designers planning placement of fill for the purpose of elevating structures and to remove sites and buildings from the mapped special flood hazard areas are referred to NFIP Technical Bulletin 10-01, *Ensuring that Structures Built on Fill In or Near Special Flood Hazard Areas Are Reasonably Safe From Flooding,* Ref. [C14]. Generally, excavation of basements into fill may expose the structure to residual risk and damage associated with flooding and saturated fill. Designers proposing basements in fill are advised excavations below the DFE are not permitted.

C2.5 SLABS-ON-GRADE AND FOOTINGS

Designers should consider extending structural fill that is used to support slabs-on-grade beyond the footprint of the foundation. Not only is access around the building improved, but the compacted fill is less likely to be disturbed during construction. The distance that the fill extends beyond the footprint of the foundation will vary as a function of the occupancy of the building.

C2.5.1 Use of Slabs-on-Grade

Guidance related to reinforcement of slabs-on-grade can be found in Ref. [C15].

Slabs installed in areas where frost depth may require turned down slabs can experience more severe erosion during a flood event. The turned downed edges can exacerbate the effects of erosion, and if considered in the design would require more reinforcement in the slab.

Slabs installed for parking or storage should be constructed with minimal reinforcement, and the edges should not be turned down so the slabs will break up during a severe flood event.

C2.6 ENCLOSURES BELOW THE DESIGN FLOOD ELEVATION

Solid perimeter walls used in crawl-space construction or to surround other areas that are designed for allowable uses may enclose areas below the DFE. Enclosures below the DFE can be used for parking, building access, and storage, provided the requirements of this standard and the authority having jurisdiction are satisfied. When a structure is located in areas not subject to high velocity wave action or high risk and where dry floodproofing is not employed, openings in enclosures below the DFE are required to permit the automatic entry and exit of floodwaters. However, the designer is advised to keep use of enclosures below the DFE to a minimum. The designer is also advised that enclosures below the DFE may result in substantially higher flood insurance premiums. Insurance implications of enclosures should be investigated.

Designers should be aware of specific situations relative to crawl-space construction. In general, in addition to flood openings (Section 2.6.1), to facilitate drainage of accumulated floodwaters the interior grade is to be at or above the exterior grade (the NFIP considers any enclosure that is below grade on all sides to be a basement, and basements are not permitted). An exception to this is the situation that may be applicable in very limited circumstances: in flood hazard areas where the depth of water above grade is less than 2 ft, a below-grade crawl space is allowed if no more than 2 ft below grade. Such below-grade crawl spaces are to be designed specifically in accordance with NFIP Technical Bulletin 11-01, *Crawlspace Construction for Buildings in Special Flood Hazard Areas,* Ref. [C16].

C2.6.1 Required Openings in Foundation Walls

Openings are required in foundation walls that enclose areas below the DFE (other than areas in nonresidential structures that are designed in accordance with Section 6 of this standard to be dry floodproofed) in order to equalize hydrostatic pressures and prevent damage or collapse.

C2.6.1.1 Openings in Breakaway Walls

This standard requires openings in breakaway walls surrounding enclosures that are located under buildings in any flood hazard area except Coastal High Hazard Areas.

C2.6.2 Design of Openings

Openings in enclosures below the DFE can be engineered or nonengineered, but should safely allow equalization of hydrostatic pressure outside and inside any enclosure below the DFE. The designer can refer to Ref. [C17] for an expanded discussion of opening requirements.

C2.6.2.1 Nonengineered Openings

The opening requirement for nonengineered openings (1 square inch of net opening area per square foot of enclosed area) is taken from the analysis described in Ref. [C17], using certain assumptions regarding the flood water rate of rise, opening coefficient of discharge, and factor of safety.

The flood opening requirement may be satisfied by use of ventilation units that commonly are used for foundation ventilation. When foundation ventilation units are installed to meet the openings requirement, such units are to be disabled in the open position in order to satisfy the requirement that flood openings allow for the automatic entry and exit of floodwaters. To determine the total open area provided, the actual open area shall be measured. Area that is obstructed or covered by a face plate, frame, grill, or other covering (other than screen) shall not be counted as open area.

C2.6.2.2 Engineered Openings

The equations contained in Section 2.6.2.2 are modified versions of an equation contained in Ref. [C17]. That reference rearranges the terms in an equation describing flow, Q, through an opening of area, A, under a head differential, H [i.e., $Q = (c)(A)\,(2gH)^{0.5}$], and adds a factor of safety, to compute the opening area required to conduct flow Q:

$$A = \{Q \times (FS)\} \div \{38.0\,(c)(p)^{0.5}\}$$

where

A = the net area of openings required (in.^{2});
Q = flow rate per square foot of opening area (gallons per minute);
FS = factor of safety, taken to be 5;
c = coefficient of discharge;
p = hydrostatic pressure due to a 1 ft head differential (pounds per square inch); and
38.0 = constant for the case where the rate of rise is 5 ft/hr.

The equation yields $A = 0.83$ square inches of opening per square foot of enclosed area, or approximately 1.0 square inch per square foot of enclosed area (the nonengineered opening requirement), using a coefficient of discharge = 0.20 and under the assumptions described.

This standard generalized the equation derived in Ref. [C17] to account for different rates of rise and factors of safety.

Discharge coefficients shown in Table 2-2 were taken from or derived from standard hydraulics texts. The coefficient represents the ratio of the actual flow through an opening divided by the ideal flow, where ideal flow is given by

$$Q = A\,(2\,g\,H)^{0.5}$$

and where Q = ideal flow (cfs), A = cross-sectional area of opening (sq ft), H = depth of flow at opening (ft), and g = gravitational constant (32.2 ft/s^2).

In the case of circular openings, hydraulics references tell us the discharge coefficient for a vertical circular, sharp-edged orifice will be approximately 0.60 under low head conditions, which are required here (maximum head difference across opening equal to 1 ft). Coefficients for the rectangular and square openings were calculated for typical opening sizes (i.e., 12 in. × 12 in. or 8 in. × 16 in., corresponding to nominal masonry unit sizes) using the ideal flow discharge relationship above and the following discharge relationship for a contracted rectangular weir:

$$Q = 3.330(L - 0.2\,H)\,H^{1.5}$$

where

Q = flow through opening in cfs;
L = horizontal length, ft; and
H = depth of flow through opening, ft.

Other shapes were assigned a coefficient of 0.30 based upon flow through a V-notch or trapezoidal weir. If the designer is certain the potential for blockage by debris is small, discharge coefficients between 0.25 and 0.60 should be used depending upon the opening shape.

The minimum dimension of the opening in any direction in the plane of the wall should be 3 in. This minimum dimension will help prevent blockage by small debris and sediment. The opening may be any shape as long as the appropriate discharge coefficients are used to reflect the shape (circular, square, rectangular, etc.)

C3.0 HIGH RISK FLOOD HAZARD AREAS

C3.1 SCOPE

The nature of the hazards listed in Section 3.1 makes the identification of High Risk Flood Hazard Areas difficult and the design and construction in such areas problematic. In addition, the state or local authority having jurisdiction may have additional requirements or may not permit any building in these areas. The intensity, spatial extent, duration, and probabilities associated with these hazards are difficult to predict, leading to uncertainties associated with the delineation and management of High Risk Flood Hazard Areas. Ref. [C4] provides general guidance for management of High Risk Flood Hazard Areas. Unfortunately, the scenic beauty of many areas in which high risk flood hazards occur attracts development interest and poses a serious challenge to floodplain managers and building officials.

The basic requirements of Section 2 apply in High Risk Flood Hazard Areas that are not identified as Coastal High Hazard Areas (V zones) and Coastal A Zones. Section 3 provides additional requirements for construction in many High Risk Flood Hazard Areas contingent upon the design and construction of protective works. However, the designer is cautioned that the same problems and uncertainties, identified above, that make the identification of High Risk Flood Hazard Areas difficult also make design and construction of protective works difficult. Construction of protective works, or reliance on existing works, to protect a structure in a High Risk Flood Hazard Area may be ill-advised, especially where uncertainties about flood hazards are great.

In High Risk Flood Hazard Areas, poor maintenance or improper operation of protective works and facilities, including pumping plants, can cause damage to or failure of those protective works and facilities, and will almost certainly result in damage or destruction of structures behind the protective works. Therefore, all protective works should include implementation of a well-conceived plan for periodic inspection, maintenance, repair, and testing.

C3.2 ALLUVIAL FAN AREAS

Alluvial fan areas represent one of the most hazardous floodplain areas. Alluvial fans are geomorphic features characterized by cone- or fan-shaped deposits of boulders, gravel, sand, and fine sediments that have been eroded from upstream watersheds, and then deposited on the adjacent valley floor. Flooding that occurs on active alluvial fans is often characterized by debris and sediment-laden flows. Channel avulsion or overbank flows can result in unconfined flows on alluvial fans where flow paths are unpredictable and subject to lateral migration. In addition, these fast-moving flows present hazards associated with erosion, debris transport and deposition, and sediment transport and deposition. Alluvial fans can be found throughout the United States, but are numerous in arid and semiarid regions.

An alluvial fan flooding hazard is indicated on a community's flood hazard map by three criteria: (a) flow path uncertainty below the hydrographic apex; (b) abrupt deposit and ensuing erosion of sediment as a stream or debris flow loses its competence to carry material eroded from a steeper, upstream source area; and (c) an environment where the combination of sediment availability, slope, and topography creates a hazardous condition for which elevation on fill will not reliably mitigate the risk.

Section 3.2 provides performance standards for construction of structures on alluvial fans. Section 3.2 does not present an exact technical representation of real world behavior; however, it is clear and precise regarding application. Design and construction in an alluvial fan should be predicated upon careful review of the physical features of an individual fan, best obtained from maps, photographs, historical flood data, soils data, and personal observation.

Extensive damages experienced on alluvial fans generally result from floods exceeding the design parameters of flood control structures or hydraulic models. Extensive flood damages can be associated with surging whose flow hydraulics may exceed structural design specifications. Damages can be particularly extensive to structures that are located directly in the flow path.

Alluvial fan flood mitigation can take two approaches, upstream storage of flood or conveyance off the developed fan. Communities may wish to reserve a storage area immediately downfan from the apex, where existing structures occupying these areas would not be reconstructed once substantially damaged. Alluvial fan communities may also choose to restrict development in reserved flood passageways. However, unless a full fan flood control project is constructed, reserved flood passageways will not assure containment of the design flood. Some combination of the two approaches may be considered. Flood hazard avoidance on alluvial fans is strongly encouraged. In some cases, the local authority should consider restriction of development on the entire alluvial fan.

The construction of elevated structures in lower risk areas of an alluvial fan will not assure complete flood protection. Floods on alluvial fans have extremely unpredictable flow paths, so the design and construction standards address only design flood conditions, including the potential for surging, channel avulsion, and debris frontal waves.

Floodproofing protection in high hazard areas near the fan apex is generally not acceptable for individual structures. Alluvial fan flood damage reduction can be achieved by constructing debris collection dams at the fan apex with maintenance plans for the removal of debris after each flood. Flood conveyance channels would normally be designed to safely transport high velocity discharge from the debris dam and downstream tributaries to a disposal location such as a large river. The U.S. Army Corps of Engineers flood protection project for the City of Los Angeles is an example of alluvial fan flood damage reduction.

Freeboard for channel or levee design to convey mud and debris flows must consider the potential for sediment deposition in the channel during the event. Freeboard in excess of 5 ft should be considered for watersheds with large boulders. Impact pressure on walls during a mudslide (i.e., mudflow) event should consider the largest diameter boulder found on the alluvial fan at the elevation traveling at velocity equal to or in excess of the peak discharge velocity.

National Flood Insurance Program (NFIP) regulations regarding development on alluvial fans are contained in the Code of Federal Regulations [44 CFR Chapter 1, Sections 60.3(c)(7) and 60.3(c)(11)], and provide alluvial fan restrictions for communities in the NFIP. Ref. [C18] discusses NFIP regulations for alluvial fans.

C3.3 FLASH FLOOD AREAS

Where structures are constructed in a flash flood area, a serious risk to human life prevails. Good examples include the Big Thompson River downstream of Estes Park, Colorado, and Gatlinburg, Tennessee. Both of these locations possess steeply sloped mountain streams with limited channel capacity, and are subject to high intensity, short duration storms. Although flood warning systems have been developed to alleviate the threat to life in these areas, they cannot completely eliminate the threat due to the quickness of the flood events.

Because of the nature of analyses of rainfall and runoff data with respect to identifying flash flood areas, designers are cautioned that such analyses should be conducted by qualified professionals.

C3.3.1 Protective Works in Flash Flood Areas

Protective works in flash flood areas should provide protection against floods more severe than the

100-year flood, to significantly reduce the threat to life. Rather than site-specific protective works, flood protection in these areas typically involves flood control reservoirs, which may be hard to justify due to their large cost, potential environmental impacts, and reservoir regulation difficulties.

C3.4 MUDSLIDE AREAS

The term "mudslide" as defined in this standard is consistent with the NFIP regulations. It is intended to include only the classes of flow that are technically recognized as mudflow and mud flood. Landslides are not covered in this standard since they are not flood related.

Mudflows are nonhomogeneous, non-Newtonian, viscous, transient hyperconcentrated sediment flood events whose fluid properties change dramatically as the flow progresses downslope. Viscous mudflow behavior is a function of the fluid matrix of water and fine sediments with a significant yield stress that must be exceeded to initiate motion. Mudflow areas are identified by irregular deposits, poorly sorted boulders, debris piles, natural levees, and large boulders transported long distances on mild slopes. Alluvial fan sediment deposits have a significant percentage of slits and clays. Damage to structures is caused primarily from frontal wave impact, mudflow deposition and inundation, boulder impact, and lateral loading.

Mud floods are hyperconcentrated sediment flow, turbulent in nature, with essentially water flow behavior. Mud floods have no yield stress, and large sediment particles will settle in a quiescent flow condition. Mud flood areas are identified by eroded channels, small levees, sorted boulders, and sediment deposits with very little fine sediments (silts and clays). Mud floods will cause damage to structures primarily from frontal wave impact, inundation, boulder impact, and lateral loading.

Mudflows are hyperconcentrated flows with a sediment concentration ranging from 45% to 55% by volume. Inundation and deposits of mud, impact of mud frontal waves, flow competence (ability to transport large boulders), and high lateral loadings during mudflows can cause structures to collapse or be moved off their foundations.

Because of the nature and complexity of analyses of mudflows and mudslides, designers are cautioned that such analyses should be conducted by qualified professionals.

Landslides may exceed 55% concentration by volume and can occur when hillsides become saturated from runoff infiltration or increased ground water levels. Slope stability analyses are required to identify potential landslide areas. In earthquake-prone areas liquefaction should be considered in the hazard risk assessment. Many factors can contribute to the initiation of landslides and/or mudslides, including vegetation removal, changes in soil moisture conditions, removal of the toe of the slope, or increased loading. Usually, areas of potential landslides have surface evidence of this potential or there is historical information where landslides have occurred in the vicinity.

C3.5 EROSION-PRONE AREAS

Historical aerial photographs, topographic maps, nautical charts, and survey data can be used to investigate erosion and estimate erosion rates. Many states already carry out such investigations and have established construction setbacks based on long-term average annual erosion rates, especially in coastal areas. Some states have identified coastal areas subject to erosion up to and during 1% chance (100-year) storms. However, erosion-prone areas could be above the floodplain. Designers should research state and local erosion zone and shoreline setback requirements. Because of the nature and complexity of analyses of flood-induced erosion and long-term erosion rates, designers are cautioned that such analyses should be conducted by qualified professionals.

C3.6 HIGH VELOCITY FLOW AREAS

Prohibiting structures in high velocity flow areas is based on the potential for severe scour, unless costly protection measures are provided to divert high velocity flows or otherwise protect the structure, Ref. [C9]. High velocity flow areas can be identified based on site location (e.g., within stream meander boundaries), historic observation of flood conditions at the site, and hydraulic analyses.

High velocity flows exhibit significant turbulence and may have standing waves. The engineering models that are used to compute water surface elevation also compute the energy grade line, which is, among other things, an approximation of the height to which waves and turbulence may rise. The energy grade lines may be considerably higher than the water surface elevation. If structures are to be constructed in high velocity flow areas, designers are advised to examine available hydraulic models to determine the computed energy grade line and should use this to represent depth in calculating flood loads.

Section 4 of this standard specifically addresses design and construction requirements for construction in Coastal High Hazard Areas and Coastal A Zones. Section 4 is appropriately applied in noncoastal areas that experience wind-driven waves that should be accounted for in determining loads and developing flood-resistant designs. Such areas include long-duration flood hazard areas found on certain lake front areas and expansive riverine floodplains.

C3.8 ICE JAM AND DEBRIS AREAS

Ref. [C19] provides additional guidance for design in icejam and ice flow areas.

One way to identify flood hazard areas subject to ice jam and debris hazards is to consider the flood depth and velocity adjacent to structure. If upstream sources of ice and/or debris are present, and if the product of the flood depth and velocity at the site exceeds some critical value, say 4 ft^2/sec, then potential ice jam and debris hazards exist. The rationale for the depth-velocity criteria is based on the concept that small objects transported at velocities up to 4 ft/sec, or larger objects transported at lower velocities, would not be expected to cause significant damage to a structure, or its foundation. However, the designer should consider this analysis as preliminary, and conduct a more detailed analysis at a given site, based on site conditions and structure design. The designer should also be careful to avoid a design that traps floating ice and debris at the structure, thereby imposing potentially large and unaccounted for loads.

Section 5.3.3.5 and commentary of ASCE 7 (Standard Ref. [1]) provides guidance on impact loads.

C4.0 COASTAL HIGH HAZARD AREAS AND COASTAL A ZONES

C4.1 SCOPE

This section defines the requirements for structures that are subject to damaging wave action in Coastal High Hazard Areas and Coastal A Zones. Coastal High Hazard Areas are those areas where conditions and water depths are sufficient to support breaking waves equal to or greater than 3 ft in height, and in areas subject to wave runup depths of 3 ft or greater. The rationale for selecting the 3 ft wave criterion is described in Ref. [C20]. The necessity for considering wave runup effects in design has been established by postflood damage investigations that found significant damage to structures not elevated above wave runup elevations. The designer is cautioned that waves less than 3 ft can cause damage to structures and should be considered even when conditions are dissimilar to those in the reference.

The Design Flood Elevation in areas subject to high velocity wave action is at the wave crest elevation or wave elevation—not at the still water level.

C4.1.1 Identification of Coastal High Hazard Areas and Coastal A Zones

The wave hazard classification employed by this standard is slightly different from that used by the National Flood Insurance Program (NFIP) in mapping flood hazard areas. NFIP floodplain mapping procedures restrict consideration of wave effects to Coastal High Hazard Areas (i.e., V zones), although wave effects may also be important in some lake and riverine areas. This standard intends for structures subject to damaging wave forces to be designed and constructed to resist those forces, regardless of location or the nature of the flooding. Hence, this standard is more restrictive than the NFIP regulations with respect to the identification of flood hazard areas subject to damaging wave action.

Designing a structure to withstand wave forces or other extreme forces unnecessarily can result in significant structural, architectural, and site planning modifications, as well as an increased cost of construction. On the other hand, failure to design for these loads when they occur may result in total or near-total loss of the structure and its contents.

Proper application of this standard may result in designing and constructing for high velocity wave forces even though a site has been designated as an A zone by the NFIP. Designers using this standard should verify if high velocity wave forces are a significant design issue at a site before following the requirements of Section 2 or Section 4. Reliance solely on the V zone and A zone designations made by the NFIP may not be sufficient, since it is possible that a structure located in an area designated by the NFIP as an A zone will be subject to damaging forces.

In order for a site to be designated "subject to high velocity wave action" by the NFIP, one of the following conditions must hold:

1. Water depths at the site must be sufficient to support 3 ft wave heights, wind or seismic forces must be capable of generating 3 ft wave heights, and

local conditions must be such that the waves can propagate over the surface of the flood waters and reach the site (see Fig. C4-1); and

2. Wave runup depths must equal or exceed 3 ft (see Fig. C4-2).

A designer can determine whether or not water depths are sufficient to support 3 ft waves using a common approximation that states the minimum still-water depth capable of supporting a 3 ft wave will be 1.28 times the breaking wave height (see Ref. [C6]). Using this approximation, a 3 ft wave will not occur in a stillwater depth less than approximately 3.8 ft. Any time stillwater flood depths exceed 3.8 ft, the potential for high velocity wave action exists.

However, when water depths at a site are assessed, calculations should be based on the stillwater flood elevations minus the eroded ground elevation expected at the site during the design flood event, not the preflood ground elevation at the site. Thus, a designer should obtain information on expected design storm and long-term erosion or calculate the erosive effects of the design flood on local soils, taking into consideration the generalized erosion that may result from water and waves, and any localized erosion due to the interaction of water and waves with the structure being designed.

The presence of a 3.8 ft stillwater flood depth at a structure site is not, by itself, sufficient to classify the area as subject to high velocity wave action. In order for that designation to be accurate, the second

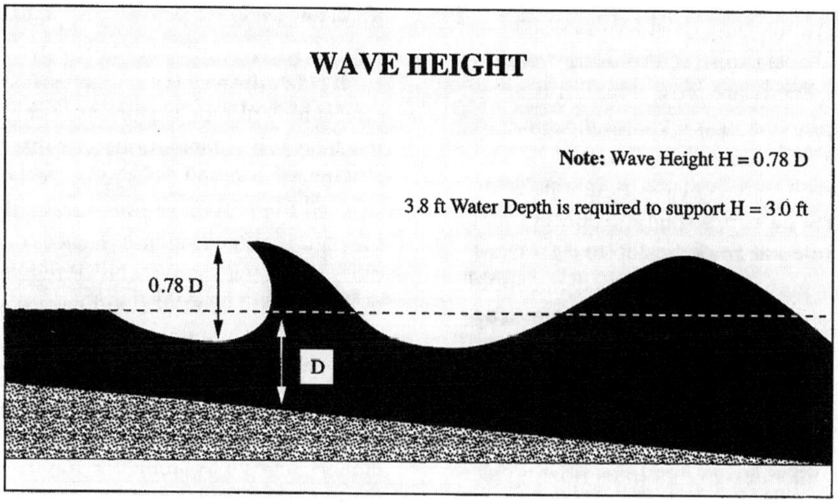

FIGURE C4-1. Definition Sketch for Still-Water Depth and Wave Height (Ref. [C21]). (Reproduced with Permission. Copyright 1977 National Academy of Sciences.)

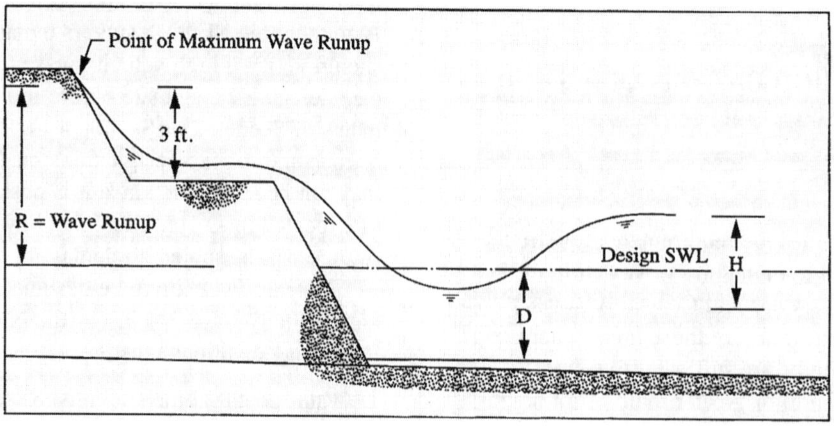

FIGURE C4-2. Definition Sketch for Wave Runup (Ref. [C6]).

condition stated above must also hold. In other words, the following must also occur: wind or seismic forces must be sufficient to generate waves equal to or greater than 3 ft in height; design flood water depths between the wave generation area and the site of interest must be sufficient to allow the waves to pass without being reduced in height below 3 ft.

In the case of wind-generated waves (the most common damaging waves), well-established procedures exist for quantifying the generation, propagation, and dissipation of waves (see Refs. [C6] and [C21]). These references will allow a designer to calculate expected wave heights and to estimate wave runup at the site of interest, thus allowing the designer to determine whether or not a structure will be subject to high velocity wave action.

This edition of the standard extends to Coastal A Zones the requirements for elevation and foundations that previously were applied only to areas subject to high velocity wave action (V Zones). The justification for increasing the design and construction requirements in Coastal A Zones is based, in part, on recent post-storm damage investigations and a critical review of the original establishment of the 3 ft breaking wave as the landward limit of the Coastal High Hazard Area. The justification and data that support the expansion are addressed in Ref. [C22] and summarized below:

1. The characteristics of flood hazards in V Zones and Coastal A Zones are largely the same—wave action, high velocity flow, erosion, and floating debris. Only the magnitudes of these hazard characteristics vary, with little diminution in magnitude in the portion of the A Zone adjacent to and inland of the V Zone.
2. In 1977, the U.S. Army Corps of Engineers (which provided much of the justification for the NFIP's original use of the 3 ft wave height) conducted an analytical study and determined that a 2.1 ft breaking wave height represented the wave height threshold for damage to conventional wood-frame walls. A more recent full-scale laboratory study, cited in Ref. [C22], determined that 1.5 ft breaking waves consistently caused failure of the bottom plate of wood-frame walls.
3. The NFIP's A Zone depth-damage functions are based largely on tens of thousands of flood insurance claims paid on buildings located in riverine A Zones. The V Zone depth-damage functions are based on thousands of claims paid on buildings located in V Zones. Examination of damage caused by Hurricane Opal in 1995 to Coastal A Zone residential structures at Pensacola Beach found that the NFIP's A Zone depth-damage functions underpredicted actual flood damages by approximately 50% to 60%. When the V Zone depth-damage functions were applied, the predicted and actual damages were within 6%.

C4.2 GENERAL

Design for wave forces requires an understanding of the mechanisms by which waves can affect structures. If of sufficient height, waves can cause damage in the following ways (see Figs. C4-3 through C4-6, taken from Ref. [C23]):

1. Waves may break against the side or underside of the structure, causing horizontal and vertical loads at least an order of magnitude higher than wind loads;
2. Waves may introduce significant drag, inertia, and other forces on structural members supporting elevated structures;
3. Waves may "peak up" or be redirected beneath an elevated structure, and create large hydrodynamic uplift forces;
4. Waves may break near the structure and run up steeply sloped ground around the structure, striking it with great force, and runup is also a factor on flatter slopes;
5. Waves may lead to instability by causing significant erosion of supporting fill beneath nonelevated structures or by causing significant erosion of soil around foundation elements supporting elevated structures; and
6. Waves can convey debris at high velocity, causing impact damage to buildings and supporting members.

If a designer finds that any of the above wave effects are likely to impact a structure, the design should take the anticipated wave loads and other loads during design flood conditions into consideration, especially for the design of the foundation and structural supports. It is not practical or economical to design most unelevated structures to resist wave forces. It is for this reason that stringent elevation and foundation requirements are imposed in this standard. Guidance for design and construction of elevated structures may be found in Refs. [C24], [C25], and [C26]. Designers are advised that the authority having jurisdiction will require documentation of the design to be signed and sealed by the designer (commonly referred to as a V Zone certification).

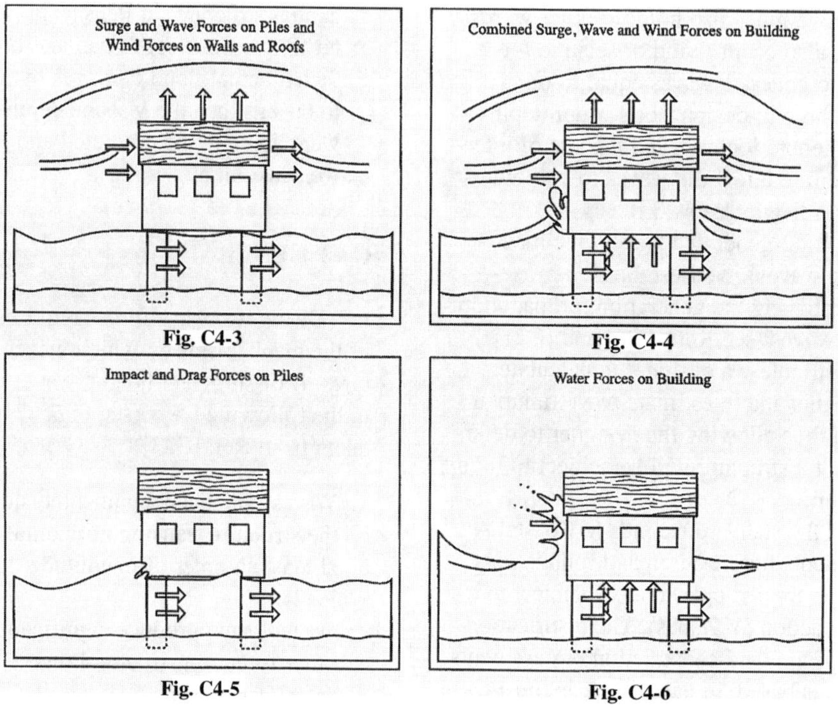

FIGURE C4-3 through 4-6. Hurricane Forces Acting on a Structure (Ref. [C23]). (Reprinted with permission. Copyright 1986 International Code Council, Inc.)

Specific guidance for calculation of wave loads is found in ASCE 7 (Standard Ref. [1]).

C4.3 SITING

Mangrove stands and sand dunes are natural barriers that reduce the landward transmission of waves and high velocity flows. The weakening or removal of mangrove stands or sand dunes as a result of construction or development activity can adversely impact properties by allowing waves and high velocity flows to penetrate further landward, exposing lands and structures to increased wave and hydrodynamic forces. No construction or development should take place that leads to a reduction of the wave and energy dissipation characteristics of sand dunes or mangrove stands.

In areas protected by sand dunes, special care must be taken during the course of any construction or development activity to prevent degradation of the structural integrity of the dune system. Man-made alteration of the dune in a Coastal High Hazard Area that can increase flood damage is prohibited by the NFIP, and alteration in an A Zone (although not prohibited by the NFIP) may also result in increased damages. Excavation of dune material should be prohibited, or, where temporary disturbance is required, replacement of such material should be required. With the exception of dune crossovers (access ways), construction and development activities should not take place on the dune.

Extreme care should be taken in siting structures in coastal areas, in recognition of the dynamic nature of many open coast, inlet, and bay/lake shorelines. Designers are cautioned that many states and communities have specific and more restrictive criteria that may affect siting, especially in environmentally sensitive areas and areas where long-term erosion rates have been evaluated.

Sites located near tidal inlets can experience higher than anticipated flood elevations and wave heights due to wave refraction and wave-current interactions. As a result, when locating a structure near a tidal inlet, it is prudent to add an additional foot of freeboard to the elevation of the lowest supporting structural member as an additional factor of safety. Similarly, the design professional should be aware of the dynamic nature of shorelines in the vicinity of tidal inlets and attempt to identify trends or patterns in shoreline movement (using historical data) that should be considered in siting structures.

Along many shorelines, average annual long-term horizontal erosion rates have been documented at 2 to 5 ft per year, or more; in rare instances, rates exceeding 20 ft per year have been reported. Construction setbacks, based on long-term shoreline change rates, have been established by many state and local governments. Therefore, the design professional should be aware of established setback regulations and seek published long-term erosion data based on historical shoreline changes. State coastal zone management agencies typically maintain or have access to historic shoreline data and should be consulted.

C4.4 ELEVATION REQUIREMENTS

The elevation of any enclosed area used for purposes other than parking, building access, and storage is set by the elevation of the lowest horizontal structural member relative to the Design Flood Elevation. The NFIP requires that the bottom of the lowest horizontal structural member be set at or above the base flood elevation.

This standard has chosen to be more restrictive, taking into account building occupancy and use, as well as orientation of the lowest horizontal structural member of the lowest floor (see Table 4-1). Note that the designer must use judgment in determining the likely direction of wave approach. In most cases, waves will approach from directly offshore, or normal to the shoreline. In some cases, the likely direction of wave approach may not be easy to determine; in such cases, the designer should use the right-hand column in Table 4-1 to determine minimum elevation.

The orientation of the lowest horizontal structural member of the lowest floor has been determined to have a significant effect on structure performance during design flood conditions, Ref. [C27]. There is substantial evidence from postevent investigations that indicate damage occurs when water strikes a structure broadside, which is the case when horizontal structural members are struck. Orienting these members to minimize the direct impact by the water will reduce flood loads being transmitted to the structure. Structures with the lowest horizontal structural member satisfying NFIP elevation requirements, but oriented perpendicular to the direction of wave approach, have sustained more damage than those structures framed with the lowest supporting member parallel to the direction of wave approach. This is most likely due to thc fact that in any given storm event, a certain fraction of wave heights will exceed predicted heights, damaging some structures built in compliance with the minimum NFIP elevation requirements. Therefore, this standard imposes additional freeboard requirements for structures with the lowest horizontal structural members perpendicular to the direction of wave approach.

C4.5.1 General

Designers should be aware that foundations in areas subject to wave action are often exposed to erosion, both storm-induced and that associated with long-term recession of the shoreline. Thus, design of foundations must account for sudden and gradual erosion. Exposing vertical foundation elements will increase the unbraced length and may require more embedment, bracing, or larger piles or some combination of all of these design modifications.

In mid-rise structures greater than about four stories, wind may be a controlling factor in the design of the structural frame, but flooding may play a significant role in the design of the lower portion of the structure. In structures of three or four stories, the effects of both winds and potential wave action will be major considerations. For taller structures, forces from wind become increasingly greater and usually determine the governing design parameters for the structure. Foundation and framing systems must be able to resist the vertical forces from dead and live loads as well as lateral forces from wind and flooding. While shear walls may be appropriate for use in mid-rise structures or higher, they must be properly engineered to resist all load combinations.

C4.5.3 Foundation Depth

Foundations for structures in areas subject to wave action should be designed and constructed in recognition of the dynamic nature of these areas. In addition to wave forces and flood-induced erosion, many areas are subject to long-term or cyclic erosion that should also be considered. Therefore, all new construction, substantial improvements, and relocations of existing structures in areas of flood hazard subject to wave action should be designed as if the structure exists in a highly erosion prone area. Pile foundations in Coastal High Hazard Areas should be embedded to no shallower than 10 ft below Mean Water Level (MWL) or into a nonerodible surface. If soil conditions prevent pile tip elevations from reaching −10 ft MWL, piles should be embedded to refusal. Note, however, that refusal may not be a sufficient depth criterion if erosion can occur to the depth of refusal strata; in such situations, foundation members should be anchored to refusal strata.

Refusal should be defined by the geotechnical investigation but can be evaluated in the field by the number of blows per inch of pile penetration. For example, in the case of wood piles—solely for the purpose of determining embedment depth and not for meeting bearing requirements—refusal can be taken to mean 4 to 5 blows per inch if driven using a standard 5,000 pound hammer dropped from a minimum 3 ft lift, or using an equivalent hammer providing a minimum of 15,000 ft-lb of energy. For prestressed concrete piles, 6 to 8 blows per inch would be a reasonable value, and for steel piles, 12 to 15 blows per inch, Ref. [C28]. Where the final pile depth is less than that called for by the design, a verification is needed to determine if the "as-built" foundation will still resist the anticipated loads during periods of erosion and scour.

Foundation depth shall consider both general and local scour effects. Thus, the designer should first obtain or calculate the expected grade elevation after wave- and/or current-induced general erosion at the site, and then consider the effects of local scour.

C4.5.4 Use of Fill

It is not the intent of this standard to prevent the construction, maintenance, or reconstruction of sand dunes.

C4.5.5 Pile Foundations

Pile embedment should be calculated based on the applied loads (see Section 1.6) and geotechnical conditions (see Section 1.5.3). Many pile foundations, particularly for residential buildings, have been installed in accordance only with past practice, and not by application of a recognized design procedure that accounts for both vertical and horizontal loads while also taking scour and erosion into account.

If soil conditions encountered during construction or installation are different from those used to calculate pile embedment ("unexpected" conditions), then pile embedment and foundation calculations should be checked for validity, and the design modified if necessary.

C4.5.5.4 Wood Piles

Construction in coastal North Carolina has used 8 × 8 square piles almost exclusively since the late 1970s, and the code has required them since the early 1980s. Field investigations of thousands of pile-supported residential structures after hurricanes in the 1990s found few apparent problems with widespread use of 8 × 8s where erosion around the foundation was not significant. However, those post-storm observations indicated that as erosion increased and/or the number of stories increased, actual or incipient foundation failures became more frequent. This suggests there is undoubtedly a limit where larger piles become appropriate.

The designer is cautioned that square timber finishes do not take treatment well where heartwood is exposed at the face. In addition, scour is known to be greater surrounding square shapes than rounded shapes.

C4.5.5.6 Concrete-Filled Steel Pipe Piles and Shells

A minimum material thickness for steel pipe piles, shells, and casings in excess of that required by service conditions may be required to resist high collapsing pressures at the lower portion of a pile. A reduced size of coarse aggregate is often necessary for filling pile shells, and should be specified accordingly. Pile shell seams should be welded unless nonwelded locked seams are proven to be capable and are approved by the authority having jurisdiction.

C4.5.5.8 Cast-in-Place Concrete Piles

Concrete mix design and pile design should be considered carefully. Some types of piles require special concrete grout or grout mixes, and special mixes may be required for concrete pumped over substantial distances.

C4.5.6.1 Pile Capacity

For sections freestanding with bottom embedment, the member can be treated as a column having an unbraced length:

$$L = H + d/12$$

where

L = unbraced length in feet;
H = height of member in feet from top of grade, plus depth of local scour, erosion, and liquefaction, to the underside of cap or other floor structure above;
d = depth of embedment, in inches, defined as the distance from top of grade minus depth of local scour and liquefaction to the point of fixity, computed conservatively as

$$d = 1.8(EI/n_h)^{1/5}$$

where

E = the modulus of elasticity of the pile (lb/in.^{2});
I = minimum moment of inertia of pile (in.^{4}); and
n_h = coefficient of horizontal subgrade reactions (lb/in.^{3}).

An appropriate effective length factor K, a function of end restraint, shall be applied to the unbraced length L to obtain an effective length $K(L)$.

In some situations the use of the coefficient of horizontal subgrade reaction may be inappropriate due to long-term creep effects. Other considerations for horizontal subgrade reactions should be made per a geotechnical investigation.

Minimum lateral resistance of an individual pile should be at least 5% of the design axial load due to inevitable eccentricities, lateral forces, and being out of plumb.

C4.5.6.2 Capacity of the Supporting Soils

The distribution of subgrade reaction on a pile depends on both the ratio of the pressure on the pile to a corresponding displacement and the flexural rigidity of the pile. The computations required to estimate these are cumbersome and subject to error. The error is increased with smaller design lateral loads and larger assumed soil diameters (diameter of subsoils centered on the pile that can be assumed available to react against a pile). In lieu of more exact analyses, the soil diameter is limited to three times the diameter of the pile.

C.4.5.6.3 Minimum Penetration

Guidance on pile penetration and installation can be found in Ref. [C29].

C4.5.6.9 Pile Splicing

Splicing of piles is difficult and, if not done properly, can compromise the structural integrity of a foundation. Given the uncertainties related to scour of erodible soils in areas subject to high-velocity wave action and in High Risk Flood Hazard Areas, timber pile splicing should be avoided. In instances where timber piles must be spliced, the splice should be below the depth of erosion and scour predicted by an erosion analysis. In coastal areas, pile splices should be below −10 ft MWL.

C4.5.7 Posts, Piers, and Columns

Column-supported foundations can be anchored against pullout through the use of rock anchors or by penetration and grouting into nonerodible material. Where penetration into a nonerodible material is chosen, the hole into the nonerodible material should be bell-shaped and filled with grout or concrete such that the lower end of the mass will be wider than the upper, and so that pullout will be resisted even with minimal shrinkage of the grout. It is highly recommended that columns be supported on piles in this area. Field experience has shown that most columns were poorly constructed and have failed.

Because of frequent difficulties in quality control of site-constructed piers, posts, and columns in residential development, piles are recommended in lieu of posts, piers, and columns. Posts, piers, and columns may not be desirable in erosion-prone or wave-prone structures.

Footings that support columns usually must be extremely large to resist overturning caused by water, wind, or seismic forces. Footing sizes can approach a practically continuous footing or mat at the bases of the columns. Large foundations will increase the potential for erosion and scour.

C4.5.7.3 Reinforced Concrete Columns

Concrete columns measuring less than 10 in. on a side or 12 in. in diameter have been found to be more susceptible to failure during flood events. While a smaller column may be adequate according to design calculations, numerous post-disaster structural assessments and field inspections have observed that smaller size columns frequently contribute to or cause structural failures due to undetermined causes.

C4.5.8 Footings, Mats, Rafts, and Slabs-on-Grade

Footings may not be desirable in erosion-prone or wave-prone structures.

C4.5.9 Grade Beams

In High Risk Flood Hazard Areas or in areas subject to high velocity wave action, the use of grade beams may increase the forces transmitted from the foundation to the structure and may increase scour around the foundation. However, they are often used in engineered structures to strengthen piling or other vertical members during the most severe flood conditions. The increased scour and increased loads should be mitigated by proper design of the pile system, grade beam, and structural frame.

C4.5.10 Bracing

Cross bracing perpendicular to the primary wave and hydrodynamic forces has been limited to rods and cables to reduce the cross-sectional area of presentment to wave forces, and because past evaluations of structure performance following major coastal storms have shown other types of cross bracing in this direction have a high rate of failure. In designing cross bracing, the designer should consider the likelihood of debris buildup against the bracing, which will increase the surface area for wave forces to act upon and increase the loads to be resisted.

In general, it is recommended the designer strive to establish a stable design free of bracing and to use bracing only to add rigidity to the design for the comfort of the occupants. Past experience has shown cross bracing often fails during a storm event and does not provide the degree of support anticipated. Knee bracing has been shown to be more reliable and is recommended as the first type of bracing to be considered.

When using steel rods for cross bracing in a highly corrosive or marine environment, the rods should be hot-dipped galvanized or fabricated from noncorrosive material.

C4.5.11 Shear Walls

Shear walls extending below the Design Flood Elevation are required to resist flood forces. Where possible, these walls should be oriented such that flood load transfer to the rest of the structure, including the foundation, is minimized. Where feasible, shear walls should be built in segments such that there are breaks in the wall that will allow flood water to flow easily around the walls.

C4.6 ENCLOSED AREAS BELOW DESIGN FLOOD ELEVATION

The NFIP requires that the use of enclosed areas below the design flood elevation be restricted to parking, access, or storage; lower areas must not be finished or used for any other purpose. No mechanical, electrical, or plumbing equipment, exclusive of risers, should be installed below the design flood elevation.

Though the NFIP places no restrictions on the size of enclosed areas below the design flood elevation, the designer is cautioned that the size of the enclosed area can affect flood insurance premiums. The designer may wish to contact an insurance agent to obtain information regarding enclosures and additional insurance premium charges.

C4.6.1 Breakaway Walls

The area below the base flood elevation may be enclosed only with breakaway elements. The breakaway feature can be accomplished either through material failure or through failure of connections holding the walls in place. Where connection failure is used, care should be taken to ensure that individual panels will not lead to unacceptable flood and debris loads being transferred to the main structure. Designers are referred to the NFIP Technical Bulletin 9-99, *Design and Construction Guidance for Breakaway Walls Below Elevated Buildings in Coastal Areas,* Ref. [C30].

C4.6.2 Openings in Breakaway Walls in Coastal A Zones

This standard requires flood openings in breakaway walls around enclosures under buildings in Coastal A Zones. Openings have long been required in all zones shown on Flood Insurance Rate Maps as A Zones. Continuing to require openings will minimize confusion by community officials, surveyors, and insurance agents.

Openings are not required in breakaway walls of enclosures under buildings in Coastal High Hazard Areas.

C5.0 MATERIALS

C5.1 GENERAL

Several sources of information are available to assist in the selection of materials for flood-resistant design and construction. These include the following references: *Flood Proofing Regulations,* Ref. [C31], and *Flood-Resistant Materials Requirements for Buildings Located in Special Flood Hazard Areas,* Ref. [C1]. Additionally, the ICC Evaluation Service, Inc., (www.icc-es.org), has a National Evaluation Services (NES) evaluation protocol entitled *Determination of Flood-Resistance Properties of Building Elements,* Ref. [C31a], to assist in the assessment of building materials, products, and assemblies with respect to the ability of such materials to resist damage due to the effects of flooding. NES protocols were developed through a different process than the ICC-ES process for development of evaluation guidelines and acceptance criteria.

Flood-damage-resistant materials are to be used below the minimum elevation specified in Table 5-1. Designers should consider using such materials above the minimum elevation in order to reduce the likelihood of damage due to wetting by waves.

C5.2.1 Metal Connectors and Fasteners

Exposed metal connectors and fasteners should be designed to a higher safety or load factor in coastal areas susceptible to salt spray to account for loss of material due to corrosion. Even when protected by a galvanized coating, corrosion has been shown to reduce the strength and effectiveness of metal connectors and fasteners in just a few years. At a minimum, building owners should undertake periodic inspection and maintenance, including painting or replacement, of connectors and fasteners. The designer should strongly consider specifying stainless steel connectors and fasteners, or nonmetallic connectors and fasteners that satisfy the requirements of this standard. Also, refer to NFIP Technical Bulletin 8-96 *Corrosion Protection for Metal Connectors in Coastal Areas,* Ref. [C32].

Applications in highly corrosive environments, such as near bodies of salt water with regular, breaking waves that are subject to salt-laden air require additional corrosion protection and, depending on the severity of the application, should

1. Be either 304 or 316 alloy stainless steel, in accordance with ASTM A276 *Standard Specification for Stainless Steel Bars and Shapes,* Ref. [C33]; or
2. Have thicker hot-dip galvanizing as provided in NFIP Technical Bulletin 8-96 *Corrosion Protection for Metal Connectors in Coastal Areas,* Ref. [C32].

C5.2.2 Structural Steel

Exposed structural steel shapes, such as W sections, tees, plates, bars, pipes, channels, and angles used for beams, columns, bracing, and lintels should be avoided in flood hazard areas, wherever possible, especially in coastal areas. Even when protected by a galvanized coating or other means, corrosion has been observed to occur rapidly.

C5.2.3 Concrete

Designers may wish to refer to *Code Requirements for Environmental Engineering Concrete Structures* (ACI 350) *and Commentary* (ACI 350R), Ref. [C34], when designing concrete structures in flood hazard areas. While the requirements may be conservative for many applications, they may be useful for certain structures that might require or benefit from increased protection.

C5.2.5 Wood and Timber

Preservative treatment of wood and timber should conform to AWPA Standards, Refs. [C35–C46].

Designers are cautioned that the growing use of wood I-joists and engineered wood floor systems does not eliminate the need for preservative treatment and proper selection of materials and components. Many of these products are not flood or moisture resistant, and may sustain unacceptable levels of permanent creep when they become wet or moist under loaded conditions, Ref. [C47]. Specification of these products for support of lowest floor systems of buildings in flood hazard areas should be matched to the expected flood conditions (depth and duration) and moisture beneath the buildings, especially if the area under the elevated building is enclosed, in which cases drainage should be considered to minimize damp conditions. Untreated or non-flood-resistant engineered wood products should not be used where those products are expected to be submerged, to be inundated, or to remain moist for long periods of time.

C6.0 DRY AND WET FLOODPROOFING

C6.1 SCOPE

Structures in certain flood hazard areas may be designed and constructed so that, together with attendant utility and sanitary facilities, areas below the minimum specified elevation are watertight with walls substantially impermeable to the passage of water and with structural components having the capability of resisting hydrostatic and hydrodynamic loads and effects of buoyancy.

For consistency with the National Flood Insurance Program (NFIP), dry floodproofing is applicable only for structures, and portions of structures, that are not used for residential purposes. The NFIP considers that residential uses include, but are not limited to, houses, dwelling units in apartments, sleeping rooms for rent or lease, areas of hospitals and medical facilities where patients are housed, and child and elder day care. For suggestions for existing, nonconforming residential structures, designers are referred to FEMA's *Homeowner's Guide to Retrofitting: Six Ways to Protect Your House from Flooding,* Ref. [C48].

Planning and design of floodproofing measures should consider the implications, including aesthetics, construction costs, alternative designs, and flood insurance premium charges, of floodproofing to an elevation higher than the minimum required elevation. Structures that are floodproofed to an additional 1 ft or more above the minimum elevations specified in the standard may, while being more costly to construct, qualify for lower federal flood insurance premium rates.

In general, because of the forces involved and the likely severe damage and consequences if floodwaters exceed the design flood elevation, floodproofing for water depths of more than 2 to 3 ft is less common. For nonconforming, existing structures, relocation and elevation measures are most likely to successfully protect flood-prone structures; however, other floodproofing measures, when properly designed for existing structures, can also provide flood protection. For new structures, floodproofing measures built into a structure and requiring no human intervention during the flood event are the most reliable and preferred method of floodproofing for all structures located in the flood hazard area. Floodproofing measures requiring human intervention inherently have a greater chance of failure.

There are many documents that provide detailed information on planning, design, and construction of these floodproofing measures. Among the more notable are the U.S. Army Corps of Engineers publication, *Flood Proofing Regulations,* Ref. [C31], and two FEMA publications, *Floodproofing Non-Residential Structures,* Ref. [C49], and *Engineering Principles and Practices for Retrofitting Flood Prone Residential Buildings,* Ref. [C50]. These documents provide information for engineers, architects, city officials, and building owners regarding all aspects of floodproofing existing structures in flood hazard areas.

NFIP regulations regarding floodproofing, contained in 44 CFR Chap. 1 subparagraphs 60.3c(3), (4), and (8), should also be consulted. These regulations are discussed in NFIP Technical Bulletin 3-93, *Non-Residential Floodproofing—Requirements and Certification for Buildings Located in Special Flood Hazard Areas,* Ref. [C51].

Design of levees and floodwalls intended to provide flood protection to an individual or group of structures are outside the scope of this standard. There are numerous engineering, nonengineering, and regulatory issues involved in the design and construction of levees and floodwalls. Therefore, caution should be exercised when considering employing a levee or floodwall as a method of providing flood protection. Local floodplain management officials, and various state and federal agencies that may have regulatory authority over levees and floodwalls, should be consulted.

C6.2 DRY FLOODPROOFING

Whenever dry floodproofing is proposed for the lowest story of a new structure, whether it be above grade, below grade, or a combination of the two, assurance must be provided that reliable flood protection will be achieved and that the structure will be substantially impermeable to the passage of floodwater, against all floods up to and including the design flood. This requires strict adherence to materials and construction requirements for dry floodproofing. Designers are advised that, as required by the NFIP, signed and sealed documentation of the design and proposed methods of construction is required.

Existing nonresidential buildings and nonresidential portions of mixed-use buildings may be dry floodproofed using retrofit methods. Extensive evaluation of the structure's ability to resist flood loads and all possible avenues of entry of water must be undertaken to ensure that retrofit methods appropriately account for flood loads and failure points. Designers should be aware that retrofit dry floodproofing should achieve the same performance expected of new structures that are dry floodproofed. Where proposed work, including the dry floodproofing work and all other work, constitutes substantial improvement, compliance with the requirements for new structures is required. Designs and proposed methods of construction are required to be accompanied by documentation that is signed and sealed by the designer.

C6.2.1 Dry Floodproofing Limitations

Dry floodproofing in residential structures, and residential portions of mixed-use structures, is not permitted because it frequently requires human action such as installing flood shields, maintaining the protective features, having an operational plan, and being able to take action within a reasonable warning time. The possible failure of the homeowner or other occupants to take such an action because of absence, lack of maintenance, or change in ownership of the home is regarded as an unacceptable risk.

Because of the nature of the loads and the chance for increased erosion and scour along solid walls, dry floodproofing measures are not allowed in structures in Coastal High Hazard Areas and Coastal A Zones. Walls below elevated structures in these areas are to meet the design requirements for breakaway walls.

The 5 ft/sec velocity restriction for floodproofing is not a requirement of the NFIP, but is used in the regulations of the U.S. Army Corps of Engineers, *Flood Proofing Regulations,* Ref. [C31] in design of structures exposed to water loads from stagnant or flowing waters. Although effective dry floodproofing can be designed for higher velocities, this is a reasonable existing limit that addresses safety of dry floodproofed structures during a flood.

C6.2.2 Dry Floodproofing Requirements

For safety, dry floodproofed buildings should not remain occupied during conditions of flooding. The requirement that at least one door be above the applicable flood elevations is intended to provide occupants that do not evacuate prior to the onset of flooding at least one access point that is above the design flood elevation.

C6.2.3 Limits on Human Intervention

In some instances, active floodproofing that requires human intervention to be effective provides an economic and appropriate means of floodproofing a structure. However, adequate warning from a credible source must be available, and strict guidelines for implementing active floodproofing must be followed (persons responsible for installing or implementing active floodproofing must be familiar with the procedures and equipment; sufficient warning time must be given to ensure the measures are put into place, etc.). Guidance on flood warnings is available in a document entitled *Automated Local Flood Warning Systems Handbook,* Ref. [C52].

C6.3 WET FLOODPROOFING LIMITATIONS

Wet floodproofing techniques are used to reduce flood damages when an enclosed area of a structure is designed to allow entry and exit of floodwaters. However, all construction materials below the Design Flood Elevation should be flood resistant and have appropriate structural strength to resist flood forces. NFIP Technical Bulletin 7-93, *Wet Floodproofing Requirements for Structures Located in Special Flood Hazard Areas,* Ref. [C5], provides information on NFIP regulations regarding floodproofing. The areas for which wet floodproofing is permitted include enclosed areas used solely for building access, parking, and storage, as well as certain types of agricultural structures. Some examples of enclosed areas that may be wet floodproofed include garages, crawl spaces, entrance foyers and stairwells, and areas beneath elevated buildings.

In some instances, active floodproofing provides an economic and appropriate means of floodproofing a structure. However, human intervention is required, and strict guidelines for implementing active floodproofing must be followed (persons responsible for installing or implementing active floodproofing must be familiar with the procedures and equipment; sufficient warning time must be given to ensure the measures are put into place, etc.). Guidance on flood warnings is available in a document entitled *Automated Local Flood Warning Systems Handbook,* Ref. [C52].

C7.0 UTILITIES

C7.1 GENERAL

To avoid damage by floodwaters up to the design flood elevation, electrical, plumbing, mechanical, HVAC, and other systems are to be located or elevated above the minimum elevations in Table 7-1. Elevated equipment should also be anchored to resist damage from high wind. Alternatively, if located below the minimum elevations in Table 7-1, those systems and components are to be specifically designed and protected to resist damage either by preventing floodwater from entering or accumulating within them or by locating the systems and components within enclosed structures that are dry floodproofed in accordance with Section 6.

Flood damage to utilities and attendant equipment may result in effluent discharge from sewer lines, contamination of potable water, fire hazards from damage to gas or electric appliances and lines, and other preventable losses. Further, damage to equipment and appliances or disruption of utility services can cause a structure to be uninhabitable after a flood, even if the structure is otherwise undamaged. The provisions of this section apply to all utilities unless specifically exempted by the authority having jurisdiction as expendable and unlikely to endanger other components or structures in the event of damage to the utility.

Protecting Building Utilities from Flood Damage: Principles and Practices for the Design and Construction of Flood Resistant Building Utility Systems (FEMA 348), Ref. [C53], outlines guidance from the National Flood Insurance Program (NFIP) and methods to achieve the performance necessary to minimize flood damage.

In addition to the requirements above, in Coastal High Hazard Areas and in Coastal A Zones, portions of electrical, plumbing, mechanical, HVAC, and other systems that extend below the design flood elevation are to be designed to resist anticipated flood loads, wave loads, and erosion and scour expected during the design flood conditions. If portions of these systems are in protective floodproofed enclosures that are attached to structures, the enclosures and the structures are to be designed for flood loads and wave loads, including loads transmitted to the structures. *Coastal Construction Manual: Principles and Practices of Planning, Siting, Designing, Constructing, and Maintaining Residential Buildings in Coastal Areas* (FEMA 55), Ref. [C24], provides more specific guidance on designing in the coastal environment, including the design of supporting exterior platforms for utilities and attendant mechanical equipment.

Where utilities and attendant mechanical equipment are allowed below the elevation in Table 7-1, a good practice is to place such installations on the most sheltered side of foundation members. In all cases, utilities and conduits, cables, switches, pipes, faucets, and other such attendant components are not to be located on or penetrate through walls designed to break away. Postflood investigations indicate that such installations adversely affect the ability of the wall to break away as designed.

C7.2 ELECTRICAL SERVICE

Installations of certain components of electrical service that are exterior to structures generally are regulated by state public service commissions and may not be subject to local requirements. Whether electrical service is distributed underground or overhead, flood damage is minimized and service can be more rapidly restored if meter bases are elevated above the design flood elevation. If elevating the meter base makes meter reading difficult, a low platform or an automated meter reading system may be appropriate. Similarly, damage is minimized if transformers, switchgear, and other exterior equipment associated with electrical distribution are located above the design flood elevation. For underground distribution, transformers are frequently pad-mounted and located below the design flood elevation. Designers should consider requesting that local utility companies elevate transformers that serve buildings in flood hazard areas. The National Electrical Code (Standard Ref. [17]) should be consulted regarding burial requirements for electrical lines.

C7.2.5 Electric Elements Installed Below Minimum Elevations

Enclosed areas below elevated buildings are allowed, provided such enclosures are built in accordance with this standard, including limitations on use. Within such enclosures, appliances and utility service, including electrical outlets and switches, are to meet the requirements of this section. If required to meet the life safety provisions of the code, or if the authority having jurisdiction deems outlets and switches to be expendable and thus allowable below the design flood elevation, then the number of such outlets and switches should be kept to the minimum necessary to serve the limited uses allowed, and the outlets and switches should be intended for wet locations. Except for some low-voltage applications, conductors are not considered expendable.

C7.3.3 Plumbing Systems Installed Below Minimum Elevations

Certain system components (such as piping) and fixtures (such as outdoor faucets and showers) may be located below the design flood elevation and meet the requirement to prevent the release of sewage into floodwaters and prevention of infiltration of floodwaters into the plumbing system (especially potable water). However, other components and fixtures (such as toilets) may not be allowed because they cannot readily meet that requirement. Backwater or backflow prevention devices may be used, for example, for floor drains, but are not to be used to allow plumbing installations below the design flood elevation unless otherwise allowed by this section. In addition, plumbing system components and fixtures should not be installed in enclosures below elevated buildings because the uses of such enclosures are to be restricted to building access, parking, and limited storage.

C7.3.4 Sanitary Systems

Designers are advised to consult with the appropriate regulatory authority, an experienced sanitation expert, or other professional regarding design and installation of on-site sanitary systems in order to provide adequate functioning without adverse health effects during flood conditions.

Postdisaster inspections have frequently discovered failures in sanitary systems that have included buoyant septic tanks, underground tanks filled with flood water and debris, interior drains that have backed up into the building from overflowing sewage disposal systems, and drain system collapse from saturated soil covering the drain field.

C7.4 MECHANICAL, HEATING, VENTILATION, AND AIR CONDITIONING SYSTEMS

Rather than place ductwork below the design flood elevation, designers are advised to consider additional elevation of the building so that all ductwork is above the design flood elevation. However, if ductwork and duct insulation are placed below the design flood elevation, the ductwork is to be designed, constructed, and installed to resist flood loads, as well as to prevent floodwaters from entering into and accumulating within the ductwork, and is to be constructed of flood-resistant materials. Duct insulation generally is not resistant to flood damage. In addition, when saturated, insulation exerts considerable load on hanger straps and connections that tend to fail, resulting in significant damage to the ductwork.

C7.4.1 Tanks

This section applies specifically to tanks that are associated with utility service to a building or that are attached to or located under a building. Other tanks are addressed in Section 9.6.

C7.5 ELEVATORS

The intent of this section is to prevent flood damage to any significant elevator component, where "significant" can be considered to include any structural components, any expensive components, or any components that are difficult (or expensive) to access or replace or restore. Thus, any significant elevator components shall be either (1) elevated above the elevation specified in Table 7-1 or (2) constructed of flood-damage-resistant materials, and be capable of resisting physical damage due to design flood conditions. The phrase "physical damage due to design flood conditions" is used in this section to indicate damage from any source during the design flood—hydrostatic, hydrodynamic, waves, flood-borne debris, sedimentation or erosion, corrosion, etc. Components below the elevations specified in Table 7-1 are not required to be flood damage resistant if (1) they are deemed not significant by the authority having jurisdiction, and (2) they are easy to access and inexpensive to replace or clean.

The NFIP Technical Bulletin 4-93, *Elevator Installation for Buildings Located in Special Flood Hazard Areas,* Ref. [C54], provides guidance related to elevator installations in flood hazard areas. Much of the following text was taken from that document.

There are two main types of elevators, hydraulic and electric traction. The hydraulic elevator consists of a car attached to the top of a hydraulic jack assembly, which normally extends below the lowest floor and is operated by hydraulic pressure. The electric traction elevator, the more common type found in structures over three or four stories, is raised by suspension ropes attached to a car and counterweights that pass over a traction sheave driven by an electric motor generally located above the elevator hoistway.

Hydraulic elevators will often require part of the assembly to be located below the Design Flood Elevation since the jack assembly is located below the lowest floor. The hydraulic pump and reservoirs, however, can easily be located up to two floors above the jack, and thus usually above the elevations specified in Table 7-1. Salt water is particularly damaging due to its corrosive nature.

Potential flood damage to elevators can be dramatically reduced when using traction elevators since the majority of the equipment is normally located above the elevator shaft and therefore not in contact with flood waters. When some equipment cannot be located above the elevations specified in Table 7-1, such as oil buffers, compensation cable, and governor rope sheaves, which usually are located at the bottom of the shaft, flood-resistant materials should be used where possible.

To minimize loss of life, injuries, and damages, a system of interlocking controls with one or more float switches in the elevator shaft, which prevents the cab from descending into floodwaters, should be installed. In addition, the majority of the electrical equipment for both elevator types, such as electrical junction boxes and circuit and control panels, should be elevated above the elevations specified in Table 7-1.

C8.0 BUILDING ACCESS

C8.1 GENERAL

Buildings in flood hazard areas, even if designed to resist flood loads, are not intended to be occupied during flooding events. Matters related to flood warning and evacuation are outside the scope of this standard.

Exterior stairs and ramps located in flood hazard areas subject to wave action or high velocity flow should be designed to transfer minimal loads to the foundation and structure supported above. Orientation parallel to the direction of flow, open rails, open risers, or designed breakaway risers, and retractable stairs are just a few methods of reducing or eliminating loads that would otherwise be transferred to the rest of the structure.

Breakaway stairs, ramps, walls, and enclosures should be designed and constructed to fail without damaging the structure to which they are attached. Breakaway components should be designed to break into small pieces to avoid becoming debris that has the potential to cause structural damage to other structures.

An alternative to traditional in-place stair design is to use retractable stairs that can be "raised" above the elevations specified in Tables 2-1 and 4-1 during periods of pending flood conditions. To accomplish this, the top of the stairway that is located at or above these elevations is hinged; the bottom of the stairway is attached to a pulley-type system so that the bottom of the stairway may be raised to or above these elevations, thus removing the stairway from the flow path of the floodwaters. These stairs should be designed in compliance with local building codes.

Foyers and enclosed or unenclosed stairways and ramps may be located under elevated buildings to provide access. If an enclosed area is designed and built to meet the applicable requirements of this standard, then the enclosure is not the lowest floor for the purpose of flood insurance rating. Designers should consider installing doors at the elevated floor level that will provide additional protection to the interior of the building from high winds and wind-driven rain in the event such enclosures break away as intended.

C9.0 MISCELLANEOUS CONSTRUCTION

C9.2 DECKS, PORCHES, AND PATIOS

In areas of flood hazard subject to waves, currents, and fast-moving waters, results of poststorm evaluations and surveys have indicated that elevated decks and porches are often founded on insufficiently embedded posts, columns, or piles. The decks or porches become undermined, transferring loads to the primary structures, causing the decks or porches to crash into the structure, or they collapse and become susceptible to wave attack, which breaks them apart, causing an increase in debris damage to the structure. Therefore, since areas of flood hazard subject to wave action or high velocity flows are often highly susceptible to erosion, adequate embedment depth is integral to the survival of the deck, porch, and patio.

Cantilevered and knee-braced decks and porches are recommended for areas subject to erosion since the structural elements (piles, columns, etc.) rely on support from the primary foundation system, which is required to be set much deeper, making damage due to erosion less likely. Where the deck is to rely on its own supporting foundation, it is recommended the foundation meet the design requirements of the primary structure.

C9.2.1 Attached Decks, Porches, and Patios

Cantilevered and knee-braced decks and porches are recommended, especially in areas subject to high velocity flows, waves, and erosion. If any portion of attached decks, porches, and patios extends below the Design Flood Elevation (DFE), then the primary structure and attachments are to be designed for the transferred loads.

C9.2.2 Detached Decks, Porches, and Patios

Where decks, porches, and patios are detached, they are to be designed and situated so as not to adversely affect the primary structure and adjacent structures through the diversion of currents and floodwaters.

C9.3 GARAGES

C9.3.1 Attached Garages and Carports

Attached garages and attached carports may be located adjacent to or under elevated buildings provided the primary structures and garages or carports are designed for the anticipated loads. If the garages or carports are designed and built to meet the applicable requirements for enclosures below the DFE, then the floor of the garage or carport is not the lowest floor for the purpose of flood insurance rating. For flood hazard areas other than Coastal High Hazard Areas, Coastal A Zones, and other High Risk Flood Hazard Areas, the requirements include flood openings, flood-resistant materials below the DFE, and elevated utilities and equipment. Enclosed garages attached to structures located below the DFE and above the lowest ground elevation in areas not subject to high velocity wave action require openings to allow balancing of hydrostatic forces. Adequate entry and exit of floodwaters is not ensured by gaps between the door segments and the garage door and the garage door jamb. In addition, opening requirements are not satisfied when human judgment and action are needed to open the garage door or other doors prior to flooding.

Below-grade areas, including garages, are not permitted in residential structures. For nonresidential structures and the nonresidential portions of mixed-use structures, garages may be designed and constructed with below-grade areas provided they are not located in Coastal High Hazard Areas, Coastal A Zones, and other High Risk Flood Hazard Areas, and provided they are floodproofed in accordance with this standard and Ref. [C51].

One of the critical elements in the floodproofing design of a below-grade garage is the point where the garage entrance ramp meets the street grade. The optimum design would be to design the entrance at or above the DFE. However, since the garage entry must often meet street grade elevations that are below the DFE, if the limitations on human intervention outlined in Section 6.2.3 are satisfied, the garage can be designed with a high-strength flood shield that can withstand the anticipated hydrostatic pressures and duration of flooding to prevent floodwaters from entering the dry floodproofed garage. In addition, a sufficient number of emergency exits should be available so that anyone in the garage will not be trapped by rising floodwaters.

C9.4 CHIMNEYS AND FIREPLACES

Chimneys and fireplaces are typically either masonry or factory-built. Similar to decks and porches, chimneys are often supported on inadequately embedded foundations that may be undermined by waves, currents, and fast-moving waters that may result in erosion and scour, thus causing the chimney to crack or separate from the structure. Therefore, adequately embedded foundations are essential to the reduction in flood damage to chimneys. It is recommended that chimneys be elevated above the design flood elevation, even in areas not subject to wave action.

C9.6 STORAGE TANKS

This section does not apply to above-ground and subsurface storage tanks that are associated with utility service to a building or that are attached to or located under a building, which are addressed in Section 7.4.1.

The factor of safety of 1.5 is a traditional value in flood-resistant design, and may differ from that in other design standards. Conflicts or differences between this standard and other applicable standards should be resolved with the authority having jurisdiction such that the performance requirements of the NFIP are equaled or exceeded, including consideration of the type and size of tank for which the factor of safety is intended.

C10.0 REFERENCES

All applicable documents referenced in the commentary are listed below with full titles and dates.

[C1] Federal Emergency Management Agency, 1993. Flood-Resistant Materials Requirements for Buildings Located in Special Flood Hazard Areas, National Flood Insurance Program Technical Bulletin 2-93. Washington, DC.

[C2] Federal Emergency Management Agency, 1995. Guide to Flood Insurance Rate Maps, FEMA 258. Washington, DC.

[C3] Federal Emergency Management Agency, 1992. Floodplain Management in the United States: An Assessment Report (Volume 2: Full Report), FIA-18. Washington, DC.

[C4] Federal Emergency Management Agency, 1987. Reducing Losses in High Risk Flood Hazard Areas: A Guidebook for Local Officials, FEMA 116. Washington, DC.

[C5] Federal Emergency Management Agency, 1993. Wet Floodproofing Requirements for Structures Located in Special Flood Hazard Areas, National Flood Insurance Program Technical Bulletin 7-93. Washington, DC.

[C6] U.S. Army Corps of Engineers, 1984. Shore Protection Manual. Washington, DC.

[C7] Fowler, Jimmy E., 1993. Coastal Scour Problems and Prediction of Maximum Scour, Technical Report CERC-93-8, U.S. Army Corps of Engineers. Washington, DC.

[C8] Simons, D.B. and F. Senturk, 1977. Sediment Transport Technology. Water Resources Publications, Fort Collins, CO.

[C9] Federal Highway Administration, 2001. Hydraulic Engineering Circular 18. Evaluating Scour at Bridges. FHWA-IP-90-017. Washington, DC.

[C10] U.S. Army Corps of Engineers, Hydrologic Engineering Center, Computer Programs, (a) HEC-1, Flood Hydrograph Package, Version 4.0, September 1990; (b) HEC-2, Water Surface Profiles, Version 4.6, May 1991; and (c) HEC-RAS, River Analysis System, Version 2.0, April 1997.

[C11] Federal Emergency Management Agency, 1993. Below-Grade Parking Requirements for Buildings Located in Special Flood Hazard Areas, National Flood Insurance Program Technical Bulletin 6-93. Washington, DC.

[C12] U.S. Army Corps of Engineers, 1994. Engineering and Design: General Design and Construction Consideration Earth and Rock-Fill Dams, Engineer Manual EM 1110-2-2300. Washington, DC.

[C13] U.S. Army Corps of Engineers, 2000. Engineering and Design: Design and Construction of Levees, Engineer Manual EM 1110-2-1913. Washington, DC.

[C14] Federal Emergency Management Agency, 2001. Ensuring That Structures Built on Fill In or Near Special Flood Hazard Areas Are Reasonably Safe from Flooding, National Flood Insurance Program Technical Bulletin 10-01. Washington, DC.

[C15] American Concrete Institute, 1996. Guide for Concrete Floor and Slab Construction, ACI 302.1.

[C16] Federal Emergency Management Agency, 2001. Crawlspace Construction for Buildings in Special Flood Hazard Areas, National Flood Insurance Program Technical Bulletin 11-01. Washington, DC.

[C17] Federal Emergency Management Agency, 1993. Openings in Foundation Walls for Buildings Located in Special Flood Hazard Areas, National Flood Insurance Program Technical Bulletin 1-93. Washington, DC.

[C18] Federal Emergency Management Agency, 1989. Alluvial Fans: Hazards and Management, FEMA 165. Washington, DC.

[C19] U.S. Army Corps of Engineers, 2002. Engineering and Design: Ice Engineering, Engineer Manual EM 1110-2-1612. Washington, DC.

[C20] U.S. Army Corps of Engineers, Galveston District, 1975. Guidelines for Identifying Coastal High Hazard Zones.

[C21] National Academy of Sciences, 1977. Methodology for Calculating Wave Action Effects Associated with Storm Surges.

[C22] Jones, Christopher P., William L. Coulbourne, and Paul Tertell, 2001. Consideration of a New Flood Hazard Zone: The Coastal A Zone, 2001 ASFPM Annual Conference.

[C23] International Code Council, Inc., 1986. Hurricane Resistant Construction Manual.

[C24] Federal Emergency Management Agency, 2000. Coastal Construction Manual: Principles and Practices of Planning, Siting, Designing, Constructing, and Maintaining Residential Buildings in Coastal Areas, FEMA 55 (3rd Edition). Washington, DC.

[C25] Federal Emergency Management Agency, 2000. Above the Flood: Elevating Your Floodprone House, FEMA 347. Washington, DC.

[C26] Federal Emergency Management Agency, 1985. Manufactured Home Installation in Flood Hazard Areas, FEMA 85. Washington, DC.

[C27] Coastal Planning & Engineering/URS Consultants, Inc., 1992. Coastal Structure Damage Methodology: Final Report. Submitted to the U.S. Army Corps of Engineers, Philadelphia District.

[C28] Bowles, J.E., 1977. Foundation Analysis and Design. McGraw Hill, Inc., New York.

[C29] American Society of Civil Engineers, 1997. Standard Guidelines for the Design and Installation of Pile Foundations, ASCE 20-96. Reston, VA.

[C30] Federal Emergency Management Agency, 1999. Design and Construction Guidance for Breakaway Walls Below Elevated Buildings in Coastal Areas, National Flood Insurance Program Technical Bulletin 9-99. Washington, DC.

[C31] U.S. Army Corps of Engineers, 1995. Flood Proofing Regulations, EP 1165-2-314. Washington, DC.

[C31a] National Evaluation Service, 2000. NES Evaluation Protocol for Determination of Flood Resistant Properties of Building Elements. Falls Church, VA.

[C32] Federal Emergency Management Agency, 1993. Corrosion Protection for Metal Connectors in Coastal Areas, National Flood Insurance Program Technical Bulletin 8-96. Washington, DC.

[C33] American Society for Testing and Materials, 1992. Standard Specification for Stainless Steel Bars and Shapes, ASTM A276.

[C34] American Concrete Institute, 1989. Environmental Engineering Concrete Structures, ACI-350.

[C35] American Wood Preservers' Association, 1999. All Timber Products—Preservative Treatment by Pressure Processes, AWPA C1-99.

[C36] American Wood Preservers' Association, 1999. Lumber, Timber, Bridge Ties and Mine Ties—Preservative Treatment by Pressure Processes, AWPA C2-99.

[C37] American Wood Preservers' Association, 1999. Plywood—Preservative Treatment by Pressure Processes, AWPA C9-99.

[C38] American Wood Preservers' Association, 1999. Piles—Preservative Treatment by Pressure Processes, AWPA C3-99.

[C39] American Wood Preservers' Association, 1999. Poles—Preservative Treatment by Pressure Processes, AWPA C4-99.

[C40] American Wood Preservers' Association, 1995. Standard for Creosote Preservative Used in Land, Fresh Water, and Marine Applications, P1/P13 95.

[C41] American Wood Preservers' Association, 1998. Standard for Creosote Solutions, AWPA P2-98.

[C42] American Wood Preservers' Association, 1999. Waterborne Preservatives, AWPA P5-99.

[C43] American Wood Preservers' Association, 1999. Standard for Oil-Borne Preservatives, AWPA P8-99.

[C44] American Wood Preservers' Association, 1998. Standard for Solvents and Formulations for Organic Preservative Systems, AWPA P9-98.

[C45] American Wood Preservers' Association, 1999. Wood for Commercial Residential Construction, Preservative Treatment, AWPA C15-99.

[C46] American Wood Preservers' Association, 1999. Materials in Marine Construction, AWPA C18-99.

[C47] Johnson, Arne P., 2003. "Creep of Wood I-Joists Exposed to Abnormally High Moisture Conditions," Practice Periodical on Structural Design and Construction, Vol. 8, No. 1, February. American Society of Civil Engineers. Reston, VA.

[C48] Federal Emergency Management Agency, 1998. Homeowner's Guide to Retrofitting: Six Ways to Protect Your House from Flooding, FEMA 312. Washington, DC.

[C49] Federal Emergency Management Agency, 1986. Floodproofing Non-Residential Structures, FEMA 102. Washington, DC.

[C50] Federal Emergency Management Agency, 1995. Engineering Principles and Practices for Retrofitting Flood Prone Residential Buildings, FEMA 259. Washington, DC.

[C51] Federal Emergency Management Agency, 1993. Non-Residential Floodproofing—Requirements and Certification for Buildings Located in Special Flood Hazard Areas, National Flood Insurance Program Technical Bulletin 3-93. Washington, DC.

[C52] National Weather Service, Office of Hydrology, 1997. Automated Local Flood Warning Systems Handbook, WSHH#2. Silver Spring, MD.

[C53] Federal Emergency Management Agency, 1999. Protecting Building Utilities from Flood Damage: Principles and Practices for the Design and Construction of Flood Resistant Building Utility Systems. FEMA 348. Washington, DC.

[C54] Federal Emergency Management Agency, 1993. Elevator Installation for Buildings Located in Special Flood Hazard Areas, National Flood Insurance Program Technical Bulletin 4-93. Washington, DC.

INDEX